T0323563

MULTIPLEX AND MULTILEVEL NETWORKS

# Multiplex and Multilevel Networks


Edited by

Stefano Battiston

*Department of Banking and Finance, University of Zurich, Switzerland*

Guido Caldarelli

*IMT, Piazza S. Francesco 19, Lucca, Italy*
*ISC, Istituto dei Sistemi Complessi, CNR, Italy*
*Catchy, Laboratory of Big Data Analysis, Roma, Italy*

Antonios Garas

*Chair of Systems Design, ETH Zurich, Switzerland*

OXFORD
UNIVERSITY PRESS

Great Clarendon Street, Oxford, OX2 6DP,
United Kingdom

Oxford University Press is a department of the University of Oxford.
It furthers the University's objective of excellence in research, scholarship,
and education by publishing worldwide. Oxford is a registered trade mark of
Oxford University Press in the UK and in certain other countries



First Edition published in 2019

Impression: 1



Published in the United States of America by Oxford University Press
198 Madison Avenue, New York, NY 10016, United States of America

British Library Cataloguing in Publication Data
Data available

Library of Congress Control Number: 2018940198

ISBN 978–0–19–880945–6

DOI: 10.1093/oso/9780198809456.001.0001

Printed and bound by
CPI Group (UK) Ltd, Croydon, CR0 4YY

# Preface

As the field of complex networks entered its maturity phase, an increasing number of researchers thought that the established methodology could deal with all cases of networked systems. However, as is usually the case in the scientific enterprise, some novel observations showed that our current knowledge is fairly limited, and network theory has still long way to go until we can make such a bold claim. The ever-increasing availability of data in fields ranging from computer science to urban systems, medicine, economics, and finance showed that networks that were usually perceived as distinct and isolated are, in reality, interacting with other networks. While this sounds like a trivial observation, it was shown that interactions of different networks can lead to unexpected results, and allow systemic vulnerabilities to emerge. Nowadays, a whole series of papers, conferences, and activities has been devoted to the analysis and modeling of multiplex and multilevel networks.

Since the early days when this novel view of complex networks first highlighted these issues, the European Commission has engaged this challenge by financing research in the area of multilevel complex systems. This book summarizes the outcome of this engagement, as the scientific contribution for each chapter is based on a large collaborating effort that was part of the MULTIPLEX project (http://www.multiplexproject.eu). This project utilized 23 distinct research teams across Europe and, from 2012 to 2016, explored this new area of research.

The starting point of the MULTIPLEX project has been the science of complex systems. In mathematical terms, the signature of complexity is the appearance of regularities at multiple scales, for example, spatial and temporal correlations between topological quantities, such as the nodes' degree. For example, in spreading phenomena such as diseases or information exchange in a population, the hubs of the contact networks between individuals take a preponderant role in the various waves of the spreading. At a higher correlation level, two-point degree correlations determine the topological mixing (i.e., assortativity) properties of the network, which may slow down or enhance such spreading phenomena. Moreover, from a dynamical perspective, a process taking place on the network might coevolve with the network itself: a feedback loop can take place between the structure of the network and what happens on it. This problem, which is hard to solve even for simple cases, becomes much more complicated by the presence of different layers at which the dynamics can operate. It is thus clear that further progress in domains dominated by multilevel networks, such as the ICT domain, will certainly benefit from an understanding of how multilevel complex systems organize and operate.

As mentioned earlier, many works have shown that networks with interactions at different levels behave in a significantly different way than when in isolation. For example, dependencies between networks may induce cascading failures and sudden collapses of the entire system, as, indeed, was observed in recent large-scale electricity blackouts. Thus, a better understanding of the structure of such systems is essential for future information technology and for improving and securing everyday life in an increasingly interconnected and interdependent world. This makes the science of complex networks particularly suitable for the exploration of the many challenges that we face today, including critical infrastructures and communication systems, as well as techno-social and socioeconomic networks.

In this book, we summarize results on the development of a mathematical, computational and algorithmic framework for the study of multilevel complex networks. These results represent a noteworthy paradigm shift, beyond which a significant progress in the understanding, prediction, control, and optimization of the dynamics and robustness of complex multilevel systems can be made. Through a combination of mathematical analysis, modeling approaches, and the use of huge heterogeneous datasets, several relevant aspects related to the topological and dynamical organization and evolution of multilevel complex networks have been addressed. Additionally, the theories, models, and algorithms produced within MULTIPLEX have been tested and validated in real-world systems of relevance in economic, technological and societal arenas.

With this book, we aim to build a guide for this fascinating novel view of complex networks, by providing to scientists, practitioners, and, most importantly, students, the basic knowledge that is necessary to pursue research in this field. Closing this short preface, we would like to thank all authors for their contributions and for their fruitful collaboration.

Stefano Battiston, Guido Caldarelli, and Antonios Garas

# Contents

# 1

# Multilayer Networks

**Sergio Gómez[1], Manlio De Domenico[1], Elisa Omodei[1], Albert Solé-Ribalta[1,2], and Alex Arenas[1]**

[1]Departament d'Enginyeria Informàtica i Matemàtiques, Universitat Rovira i Virgili, Tarragona (Catalonia)
[2]Currently at Internet Interdisciplinary Institute, Universitat Oberta de Catalunya, Castelldefels, Barcelona (Catalonia)

## 1.1 Multilayer, multiplex, and interconnected networks

The development of *network science* has provided researchers and practitioners with the necessary theory and tools for the analysis of complex networks (see, e.g., [38, 210, 88, 15]). After an initial phase in which the focus was on simple and static networks, the field has evolved in the last years toward the consideration of more involved and dynamic topological structures. Thus, the literature is now full of references to evolving networks, interdependent networks, multilayer networks, multiplex networks, simplicial complexes, and hypergraphs. Here, we are going to introduce interconnected multilayer networks, analyzing them from just a few fronts: types of multilayer networks, their mathematical description, the dynamics of random walkers, and the centrality (versatility) of nodes. This selection emphasizes how the interconnected multilayer structure differs from that of the standard single-layer networks, and its influence on dynamics on top of them. For a broader review of this field, see Ref. [37].

Multilayer networks appear naturally in real data when we realize that, in many cases, the relationships (links) between the elements (nodes) can be of different kinds. For example, people are connected through friendship, family, or work relations. We may represent this structure with a network formed by three layers, one for each type of relationship, and with the same nodes repeated in all layers. This multilayer network allows for an explicit consideration of the different characteristics the dynamics may have in each layer, and the interactions between them. For instance, confidential information for a company may flow easily within the work layer, should have difficulties in jumping

Gómez, S., De Domenico, M., Omodei, E., Solé-Ribalta, A., and Arenas, A., "Multilayer Networks" in *Multiplex and Multilevel Networks*, edited by Battiston, S., Caldarelli, G., and Garas, A. © Oxford University Press 2019.
DOI: 10.1093/oso/9780198809456.003.0001

to the other layers, and could spread slowly to family or friends due to lack of interest. Thus, we have different behaviors in each layer, and interactions between them, which result in more realistic but, at the same time, more complex dynamics.

It is important to discuss the difference between the topological structure which represents the core of this study, namely *interconnected multilayer networks* [200, 119, 236, 72, 277, 122, 253], and other multilayer structures which have been named *multiplexes* in the past and have been the subject of recent studies [170, 213, 33, 27, 274]. Note that interconnected multilayer networks are not simply a special case of or equivalent to interdependent networks [106]: in multilayer systems, many or even all the nodes have a counterpart in each layer, so one can associate a vector of states to each node. This feature has no counterpart in *interdependent networks*, which were conceived as interconnected communities within a single, larger network [48, 77]. In fact, interdependent networks are characterized by having different types of nodes, instead of links.

Historically, the term *multiplex* has been adopted to indicate the presence of more than one relationship between the same actors of a social network [221]. This type of network is well understood in terms of "coloring" (or labeling) the edges corresponding to interactions of different nature. For instance, the same individual might have connections with other individuals based on financial interests (indicated by, say, a red line) and connections with the same or different individuals based on friendship (indicated by, say, a blue line). This type of network is represented by a *noninterconnected multiplex*.

Conversely, in other real-world systems, like the transportation network of a city, the same geographical position can be part, for instance, of the network of subway or the network of bus routes, simultaneously. In this specific case, an edge-colored graph would not capture the full structure of the network, because it is missing information about the cost to *move* from the subway network to the bus route. This cost can be economic

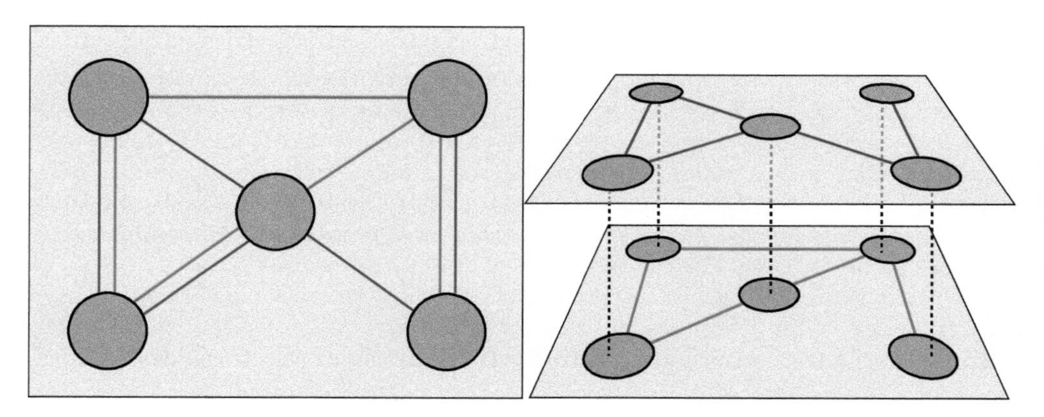

**Figure 1.1** *Edge-colored versus interconnected multilayer networks. (Left) Edge-colored graph representing two different types of interactions (purple, and green) between five actors. (Right) An interconnected multiplex representing the same actors exhibiting the same relationships but on different levels which are separated by a cost (dotted vertical lines) to move from one layer to the other.*

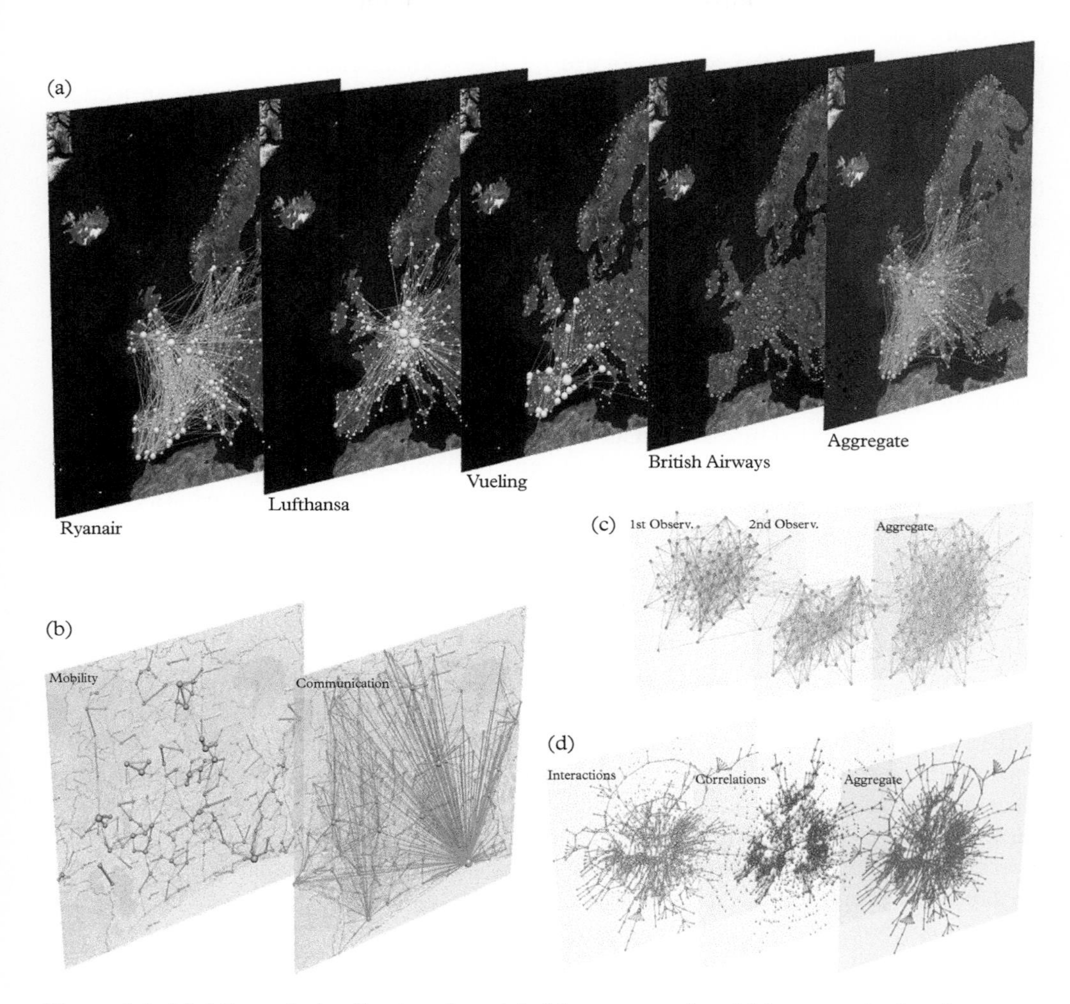


**Figure 1.2** *Multilayered visualization of empirical interconnected multiplex networks. Interlayer connections not shown for simplicity. (a) Flight routes operated by different air companies between European airports [56]. (b) Mobility and the communication networks of sub-prefectures in the Ivory Coast, built from mobile phone calls data [233, 176]. (c) Two different observations (separated by 3 weeks) of one ant colony [36]. (d) Interaction network (left) of genes in* Saccharomyces cerevisiae, *obtained through synthetic genetic array methodology, and correlation-based network (middle) connecting genes with similar genetic interaction profiles [65]. We show in each layer the largest connected component, where only pairs with high genetic interaction scores and highly correlated genetic profiles are considered. The resultant aggregated network (right) is shown, to highlight the information loss. Visualization made with MuxViz [71].*

or might account for the time required to physically commute between the two layers. Therefore, the interconnected multilayer topology presented in this section provides a better representation of the system. Figure 1.1 shows an illustration of an edge-colored graph (left) and an interconnected multiplex (right). It is evident that a simple projection of the latter—mathematically equivalent to summing up the corresponding adjacency matrices—would provide a network where the information about the colors is lost. On the other hand, an edge-colored graph cannot account for interconnections, keeping irreconcilable the two structures in Figure 1.1, which should be used to represent very different networked systems.

For further details about the classification of such multilayer networks, we refer to Ref. [149] and references therein.

A real-world example of a multiplex network is provided by the transportation network of a city, where the same geographical position can be part, for instance, of the network of a subway or the network of bus routes, simultaneously. We show in Figure 1.2(a) the case of flight routes operated by different air companies between European airports. In other examples, layers encode the human mobility and the mobile communication networks of different geographical areas (Figure 1.2(b)), or physical contacts over time between ants in a colony (Figure 1.2(c)). In biological systems, such as genetic networks, two genes might exhibit different interactions (e.g., allelic or nonallelic) or be related because of their chemical interactions or their functional roles (Figure 1.2(d)).

## 1.2 Mathematical description of multilayer networks

A first step in the analysis of multilayer networks is the development of an adequate mathematical framework. While most single-layer networks can be described with adjacency and/or weights matrices, multilayer networks have to account for intralayer edges, interlayer edges, and the possibility of having replicas of the same node in several layers. Depending on the problem at hand, the use of a set of adjacency layers, one per layer, is enough to describe the system [27]. However, this option makes sense only for edge-colored graphs, not for interconnected multilayer networks. A more powerful approach consists in describing the multilayer networks with *supra-adjacency* matrices [119, 149], which are block matrices formed by diagonal blocks to describe the intralayer adjacency matrices, and off-diagonal blocks for the interlayer connections between every pair of layers. They represent faithfully all the edges in any multilayer network, but have problems when nodes in different layers represent the same "real" node. The most general framework is provided by *tensor algebra* [72].

### 1.2.1 Tensorial formalism

Edge-colored graphs can be represented by a set of adjacency matrices [56, 213, 33, 27]. However, standard matrices, used to represent networks, are inherently limited in the complexity of the relationships that they can capture, that is, they do not represent a suitable framework in the case of interconnected multiplexes. This is the case of increasingly complicated types of relationships—which can also change in time—between nodes. Such a level of complexity can be characterized by considering tensors and higher-order algebras [72].

A great advantage of tensor formalism also relies on its compactness. An adjacency tensor can be written using a more compact notation that is very useful for the generalization to *multilayer tensor* networks. In this notation, a row vector $\mathbf{a} \in \mathbb{R}^N$ is given by a covariant vector $a_\alpha$ ($\alpha = 1, \ldots, N$), and the corresponding contravariant vector $a^\alpha$ (i.e., its dual vector) is a column vector in Euclidean space. A canonical vector is assigned to each node, and the corresponding interconnected multilayer network is represented by a rank-4 adjacency tensor.

However, in the majority of applications, it is not necessary to perform calculations using canonical vectors and tensors explicitly. Consequently, a classical single-layer network represented by a rank-2 mixed adjacency tensor $W^\alpha_\beta$ [72] can be simply indicated by $W^i_j$, where the "abuse of notation" consists in interpreting the indices $i$ and $j$ as nodes, and $W^i_j$ indicates the intensity of the relationship between them. Hence, $W^i_j$ represents the well-known adjacency matrix of a graph, and the classical notation for the weight $w_{ij}$ of the link between $i$ and $j$ corresponds to $W^i_j$. The "abuse of notation" also consists in treating $W^i_j$ as a rank-2 tensor, although it explicitly indicates the entry of a matrix, while keeping the algebraic rules governing covariant and contravariant tensors. This "abuse of notation" dramatically reduces the complexity of some tensorial equations, although it is worth remarking that it should be used only when calculations do not involve canonical tensors explicitly.

To distinguish simple networks from the more complicated situations (e.g., interconnected multiplex networks) that we use in this paper, we will use the term *monoplex networks* to describe such standard networks, which are time independent and possess only a single type of edge to connect its nodes.

In general, there might be several types of relationships between pairs of nodes and a more general system represented as a multilayer object—in which each type of relationship is encompassed in a single *layer* $\alpha$ ($\alpha = 1, 2, \ldots, L$) of a system—is required. Note that $\alpha$ no longer has the same meaning of the index in the adjacency tensor discussed above. To avoid confusion, in the following we refer to nodes with Latin letters and to layers with Greek letters, allowing us to distinguish indices that correspond to nodes from those that correspond to layers in tensorial equations.

We use an *intralayer adjacency tensor* for the second-order tensor $W^i_j(\alpha)$ that indicates the relationships between nodes within the *same* layer $\alpha$. We take into account the possibility that a node $i$ from layer $\alpha$ can be connected to any other node $j$ in any other layer $\beta$. To encode information about relationships that incorporate multiple layers, we introduce the second-order *interlayer adjacency tensor* $C^i_j(\alpha\beta)$. Note that $C^i_j(\alpha\alpha) = W^i_j(\alpha)$.

It has been shown that the mathematical object accounting for the whole interconnected multilayer structure is given by a fourth-order (i.e., rank-4) *multilayer adjacency tensor* $M^{i\alpha}_{j\beta}$. This tensor might be simply thought as a higher-order matrix with four indices. It is the direct generalization of the adjacency matrix in the case of monoplexes, encoding the intensity of the relationship (which may not be symmetric) between a node

$i$ in layer $\alpha$ and a node $j$ in layer $\beta$ [72]. This object is very general and can be used to represent structures where an actor is present in some layers but not in all of them. This is the case, for instance, when considering a network of online social relationships, of an individual with an account on Facebook but not on Twitter. The algebra still holds for these situations without any formal modification. In fact, one simply introduces "empty nodes" and assigns the value 0 to the associated edges, although the calculations of network diagnostics should carefully account for the presence of such nodes (e.g., for a proper normalization) [72].

Often, to reduce the notational complexity in the tensorial equations, the Einstein summation convention is adopted. It is applied to repeated indexes in operations that involve tensors. For example, we use this convention in the left-hand sides of the following equations:

$$A_i^i = \sum_{i=1}^{N} A_i^i,$$

$$A_j^i B_i^j = \sum_{i=1}^{N}\sum_{j=1}^{N} A_j^i B_i^j,$$

$$A_{j\beta}^{i\alpha} B_{i\gamma}^{k\beta} = \sum_{i=1}^{N}\sum_{\beta=1}^{L} A_{j\beta}^{i\alpha} B_{i\gamma}^{k\beta},$$

whose right-hand sides include the summation signs explicitly. It is straightforward to use this convention for the product of any number of tensors of any order. In the following, we will use the $t$th power of rank-4 tensors, defined by multiple tensor multiplications:

$$(A^t)_{j\beta}^{i\alpha} = (A)_{j_1\beta_1}^{i\alpha} (A)_{j_2\beta_2}^{j_1\beta_1} \cdots (A)_{j\beta}^{j_{t-1}\beta_{t-1}}. \tag{1.1}$$

Using repeated indexes, where one index is a subscript and the other is a superscript, is equivalent to performing a tensorial operation known as a *contraction*. Note that one should be very careful in performing tensorial calculations. For instance, using traditional notation, the product $a^i b^j$ would be a number, that is, the product of the components of two vectors. However, in our formulation, the same calculation denotes a Kronecker product between two vectors, resulting in a rank-2 tensor, that is, a matrix.

An interesting network that can be derived from the interconnected structure is the aggregated network, where the edges between two actors are summed up across all layers. The superposition of the different layers is equivalent to summing up the adjacency tensor of each layer. The corresponding aggregated network $G_j^i$ is a monoplex and is obtained by contracting the layer indexes of the multilayer adjacency tensor, that is, $G_j^i = M_{j\alpha}^{i\alpha}$. This aggregation loses the information about inter-layer connections. If such information is important for the application of interest, then the tensor should be contracted with the 1-tensor $u_\alpha^\beta$ (the rank-2 tensor with all components equal to 1), that is, $\bar{G}_j^i = M_{j\beta}^{i\alpha} u_\alpha^\beta$.

This formalism is extremely useful for showing how topological descriptors of interconnected networks differ from the ones corresponding to their aggregated graphs [72, 66]. Moreover, it is particularly suitable for performing compact calculations.

As a representative example, let us consider the number of paths of length 2 from a node in a certain layer to any other node in any other layer of the system. Taking advantage of the extended algebra, it is straightforward to show that the resulting rank-4 tensor accounting for such paths is given by $H^{i\alpha}_{j\beta} = M^{i\alpha}_{k\gamma} M^{k\gamma}_{j\beta}$. If only the number of paths between any pair of nodes is required, regardless of the layer, then the corresponding rank-2 tensor of paths is simply obtained by contracting with the 1-tensor $u^{\beta}_{\alpha}$, that is, $X^i_j = H^{i\alpha}_{j\beta} u^{\beta}_{\alpha}$. Conversely, in the case of the aggregate, we first contract the multilayer adjacency tensor to obtain the aggregation $\mathcal{J}^i_j = M^{i\alpha}_{j\beta} u^{\beta}_{\alpha}$, where interlayer connections are included as self-loops, and then square the resulting tensor to obtain $Y^i_j = \mathcal{J}^i_k \mathcal{J}^k_j$. Of course, a similar argument can be used to calculate the number of longer paths. From these tensorial equations, it is evident that the aggregated graph cannot be considered, in general, a good proxy of the interconnected topology.

Summarizing, the tensorial formulation provides a suitable framework for several real-world networked systems, from transportation networks to social ones. It is also worth noting that special cases of multilayer adjacency tensors are time-dependent (i.e., "temporal") networks [72, 149]. More specifically, in the case of social sciences, the multilayer adjacency tensor can be used, for instance, to model the structural changes of a social network over time, or to define the topology of actors involved in several different levels of relationships and for whom it is indispensable to define an interconnection between such levels. For these networked systems, it is desirable to adopt descriptors (e.g., clustering coefficient, modularity, etc.) that are the natural extension of their well-known counterparts in monoplex networks.

### 1.2.2 Tensorial nature of adjacency tensors

Although we have already shown in Ref. [72] the advantages of using the tensor formalism to deal with multilayer networks, the assignment of the indices as covariant or contravariant may seem arbitrary. The problem arises from the absence of natural basis transformations which could guide us in this decision. The idea is that, if we perform a change of basis governed by a matrix $Q^{\alpha}_{\beta}$, each contravariant index of any tensor is transformed using $Q$, while covariant indices change with $Q^{-1}$, the inverse of $Q$. Thus, an object with three indexes which transforms with two $Q$ and one $Q^{-1}$ is bounded to be 1-covariant and 2-contravariant. However, these transformations are not the origin but the consequence of the "meaning" of the object. For example, inner products, metric tensors, and symplectic forms must be 2-covariant, since they are bilinear functions which assign two vectors to a number, while linear transformations are 1-covariant and 1-contravariant because they have to convert a vector (or 1-form) in another vector (or 1-form).

In the case of monoplex networks, the adjacency tensor may be viewed as a linear transformation which, given a vector (or 1-form) representing a node, returns the set of their adjacent nodes. Thus, the only acceptable representation for the monoplex adjacency object is a 1-covariant and 1-contravariant tensor. Likewise, the multilayer adjacency tensor transforms a node in one layer into the set of adjacent nodes, keeping also the information of which layer they belong to; thus, a 2-covariant and 2-contravariant tensor is needed.

Once we know the order of the adjacency tensor, its transformation under a change of coordinates is completely determined. First, we show how this works for a single-layer network and, afterwards, for a full multilayer network.

By following Ref. [72], the adjacency tensor $W^{\alpha}_{\beta}$ of a network can be represented as a linear combination of tensors of the canonical basis by

$$W^{\alpha}_{\beta} = \sum_{i,j=1}^{N} w_{ij} e^{\alpha}(i) e_{\beta}(j) = \sum_{i,j=1}^{N} w_{ij} E^{\alpha}_{\beta}(ij), \tag{1.2}$$

where $E^{\alpha}_{\beta}(ij) \in \mathbb{R}^{N\times N}$ indicates the tensor of the canonical basis corresponding to the tensorial product of the canonical vectors $\mathbf{e}(i)$ and $\mathbf{e}^{\dagger}(j)$ (defined in $\mathbb{R}^{N}$) assigned to nodes $i(e^{\alpha}(i))$ and $j(e_{\beta}(j))$, respectively.

Let

$$Q^{\alpha}_{\beta} = \sum_{i=1}^{N} e'^{\alpha}(i) e_{\beta}(i) \tag{1.3}$$

be the change of basis tensor which transforms the basis vector set $\{e^{\alpha}(i), i = 1, \dots, N\}$ into a second set $\{e'^{\alpha}(i), i = 1, \dots, N\}$. Here, $Q^{\alpha}_{\beta}$ is expressed in terms of the basis vectors from both bases, and it is straightforward to show that $e'^{\alpha}(i) = Q^{\alpha}_{\beta} e^{\beta}(i)$ and $e'_{\beta}(i) = e_{\alpha}(j)(Q^{-1})^{\alpha}_{\beta}$. By remarking that a change of basis should not affect the intensity of the relationship between nodes $n_i$ and $n_j$, by following the above prescription, we obtain

$$\begin{aligned} W'^{\gamma}_{\delta} &= \sum_{i,j=1}^{N} w_{ij} e'^{\gamma}(i) e'_{\delta}(j) = \sum_{i,j=1}^{N} w_{ij} Q^{\gamma}_{\alpha} e^{\alpha}(i) e_{\beta}(j) (Q^{-1})^{\beta}_{\delta} \\ &= Q^{\gamma}_{\alpha} \left[ \sum_{i,j=1}^{N} w_{ij} e^{\alpha}(i) e_{\beta}(j) \right] (Q^{-1})^{\beta}_{\delta} = Q^{\gamma}_{\alpha} W^{\alpha}_{\beta} (Q^{-1})^{\beta}_{\delta}, \end{aligned} \tag{1.4}$$

providing the desired tensor transformation law.

In the following, we use the same notation as in Ref. [72], to avoid confusion. In the same spirit, we introduce the vectors $e^{\tilde{\gamma}}(k)$ $(\tilde{\gamma}, k = 1, \dots, L)$ of the canonical basis in the space $\mathbb{R}^{L}$, where the Greek index indicates the components of the vector, and the Latin index indicates the $k$th canonical vector. Therefore, it is straightforward to build the second-order tensors $E^{\tilde{\gamma}}_{\tilde{\delta}}(hk) = e^{\tilde{\gamma}}(h)e_{\tilde{\delta}}(k)$ representing the canonical basis of the space $\mathbb{R}^{L\times L}$.

The representation of the multilayer object $M^{\alpha\tilde{\gamma}}_{\beta\tilde{\delta}}$ in terms of the Kronecker product of canonical vectors is given by [72]

$$M^{\alpha\tilde{\gamma}}_{\beta\tilde{\delta}} = \sum_{i,j=1}^{N}\sum_{h,k=1}^{L} w_{ij}(hk)e^{\alpha}(i)e_{\beta}(j)e^{\tilde{\gamma}}(h)e_{\tilde{\delta}}(k). \tag{1.5}$$

Proceeding as in the case of a single-layer network, we obtain

$$\begin{aligned} M'^{\alpha\tilde{\gamma}}_{\beta\tilde{\delta}} &= \sum_{i,j=1}^{N}\sum_{h,k=1}^{L} w_{ij}(hk)Q^{\alpha}_{\rho}e^{\rho}(i)(Q^{-1})^{\sigma}_{\beta}e_{\sigma}(j)\tilde{Q}^{\tilde{\gamma}}_{\tilde{\phi}}e^{\tilde{\phi}}(h)(\tilde{Q}^{-1})^{\tilde{\epsilon}}_{\tilde{\delta}}e_{\tilde{\epsilon}}(k) \\ &= Q^{\alpha}_{\rho}\tilde{Q}^{\tilde{\gamma}}_{\tilde{\phi}}M^{\rho\tilde{\phi}}_{\sigma\tilde{\epsilon}}(Q^{-1})^{\sigma}_{\beta}(\tilde{Q}^{-1})^{\tilde{\epsilon}}_{\tilde{\delta}}, \end{aligned} \tag{1.6}$$

providing the desired transformation law of the multilayer adjacency tensor under a change of coordinates.

### 1.2.3 Eigenvalue problem with tensors

The eigenvalue problem for a rank-2 tensor, that is, a standard matrix, is defined by $W^{i}_{j}v_i = \lambda v_j$. The extension of this problem to rank-4 tensors leads to the equation

$$M^{i\alpha}_{j\beta}V_{i\alpha} = \lambda V_{j\beta}. \tag{1.7}$$

To solve this problem, it is worth noting that any tensor can be *unfolded* to lower-rank tensors [154]. For instance, a rank-2 tensor like $W^{i}_{j}$, with $N^2$ components, can be flattened to a vector $w_k$ with $N^2$ components. In the case of the rank-4 multilayer adjacency tensor $M^{i\alpha}_{j\beta}$, although any unfolding is allowed, it is particularly useful for some applications to choose the ones flattening to a squared rank-2 tensor $\tilde{M}^{k}_{l}$ with $NL \times NL$ components, where $L$ indicates the number of layers [119]. In fact, this unfolding produces as many block adjacency matrices, named *supra-adjacency matrices* in some applications [119, 149, 73, 66], as the number of permutations of diagonal blocks of size $N^2$, that is, $L!$. However, such unfoldings do not alter the spectral properties of the resulting supra-matrix and can be used to solve the eigenvalue problem for rank-4 tensors. In fact, the solution of the eigenvalue problem

$$\tilde{M}^{k}_{l}\tilde{v}_k = \tilde{\lambda}_1\tilde{v}_l, \tag{1.8}$$

is a *supra-vector* with $NL$ components which corresponds to the unfolding of the eigentensor $V_{i\alpha}$. We will make use of this eigenvalue formalism for tensors in Section 1.4.

## 1.3 Random walks in multiplex networks

Random walks constitute one of the simplest dynamics one can define on top of graphs or complex networks [311, 290, 180]. They can be used to approximate other types of diffusion [61, 210]. Random walks on monoplex networks [61, 214, 210] have attracted considerable interest because they are both important and easy to interpret. They have yielded important insights into a huge variety of applications and can be studied analytically. For example, random walks have been used to rank Web pages [46] and sports teams [55], optimize searches [304], investigate the efficiency of network navigation [313, 68], characterize cyclic structures in networks [247], and coarse-grain networks to illuminate mesoscale features such as community structure [115, 242, 167].

In a random walk process, the walker is initially positioned in any node and then starts to navigate the network, following the available edges. At each step, the edge is selected at random between the outgoing links of the current node, hence the name "random walker." If the network is undirected and connected, all nodes have a nonnull probability of being visited, and these probabilities are proportional to the degrees of the nodes (in the limit of paths of infinite length). The analysis of random walks for specific complex network topologies, such as networks with power-law degree distributions or small-world architectures, has revealed the different ways in which the networks are explored [214, 313].

When the network is structured in layers, the navigation of the random walker is formed by two kinds of movements: intralayer steps, in which the walker jumps between nodes within the same layer, and interlayer steps, where the walker switches from one layer to another one. In what follows, we will consider multiplex networks, the particular case of general multilayer networks in which the same nodes are present in all layers, and interlayer connections appear only between the different instances of the same node (see Figure 1.3). The multiplex networks can be weighted, thus converting the random walks in biased random walks, as we will specify below. The analysis of random walks in multiplex networks we are going to describe can be found in [73].

Given a multiplex network with $N$ nodes in each of the $L$ layers, we use $W_{ij}^{(\alpha)}$ to indicate the weighted intralayer connection between two vertices $i$ and $j$ in layer $\alpha$, where Latin letters refer to vertices ($i,j = 1,2,...,N$), and Greek letters indicate layers ($\alpha = 1,2,...,L$). Similarly, $D_{(i)}^{\alpha\beta}$ denotes the weight of switching from layer $\alpha$ to layer $\beta$ when located in a vertex $i$. Without loss of generality, we may suppose that $W_{ii}^{(\alpha)} = 0$ for all nodes $i$, since these self-loops can be accounted for in the terms $D_{(i)}^{\alpha\alpha}$. The corresponding weighted $NL \times NL$ supra-adjacency matrix becomes[1]

$$\mathcal{A} = \begin{pmatrix} \mathbf{D}^{11} + \mathbf{W}^{(1)} & \mathbf{D}^{12} & \dots & \mathbf{D}^{1L} \\ \mathbf{D}^{21} & \mathbf{D}^{22} + \mathbf{W}^{(2)} & \dots & \mathbf{D}^{2L} \\ \vdots & \vdots & \ddots & \vdots \\ \mathbf{D}^{L1} & \mathbf{D}^{L2} & \dots & \mathbf{D}^{LL} + \mathbf{W}^{(L)} \end{pmatrix}, \tag{1.9}$$

[1] In this section, we (partly) adopt the supra-adjacency formalism and notation instead of the tensorial one in order to emphasize the difference in role between intralayer and interlayer links.

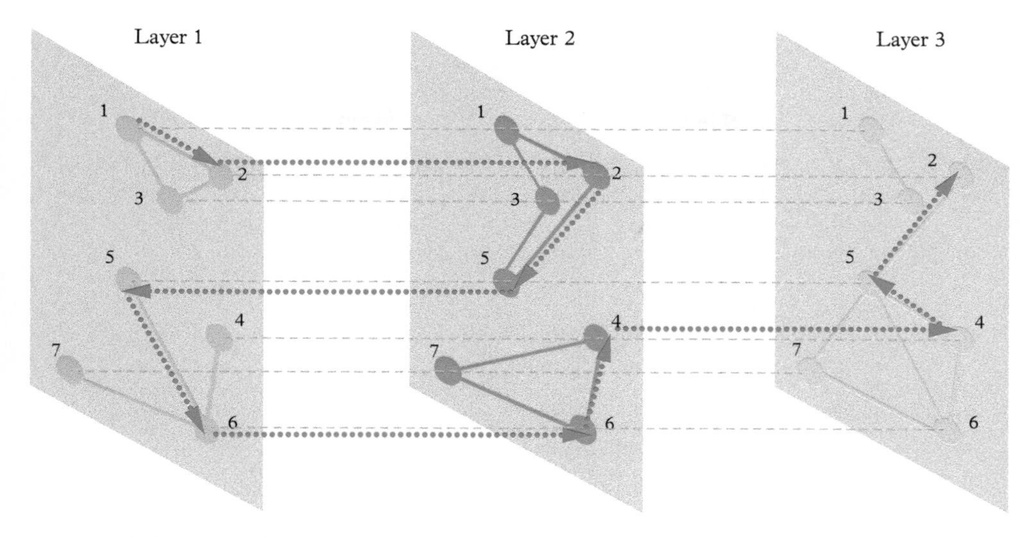


**Figure 1.3** *Example of the navigation on a multiplex network. Path (dotted pink line) of a random walker through the multiplex structure. Note how the walker is able to visit disconnected components. In this example, there are no intralayer links between Layers 1 and 3.*

where we have used boldface to represent the matrices $\mathbf{W}^{(\alpha)}$ (intralayer weights of layer $\alpha$) and $\mathbf{D}^{\alpha\beta}$ (a diagonal matrix with the weights of switching from layer $\alpha$ to layer $\beta$). A commonly studied particular case is the one in which the switching-layer weight is the same for all nodes, so that $\mathbf{D}^{\alpha\beta} = D^{\alpha\beta}\mathbf{I}$, where $\mathbf{I}$ is the $N \times N$ identity matrix:

$$\mathcal{A} = \begin{pmatrix} D^{11}\mathbf{I} + \mathbf{W}^{(1)} & D^{12}\mathbf{I} & \dots & D^{1L}\mathbf{I} \\ D^{21}\mathbf{I} & D^{22}\mathbf{I} + \mathbf{W}^{(2)} & \dots & D^{2L}\mathbf{I} \\ \vdots & \vdots & \ddots & \vdots \\ D^{L1}\mathbf{I} & D^{L2}\mathbf{I} & \dots & D^{LL}\mathbf{I} + \mathbf{W}^{(L)} \end{pmatrix}. \tag{1.10}$$

It is advantageous to distinguish between the strength $s_{i\alpha} = \sum_j W_{ij}^{(\alpha)}$ of a node $i$ with respect to its connections with other nodes $j$ in the same layer $\alpha$, and the strength $S_{i\alpha} = \sum_\beta D_{(i)}^{\alpha\beta}$ of the same vertex with respect to connections to its counterparts in different layers.

## 1.3.1 Navigation on a multiplex network

The description of the random walk navigation on a multiplex network is completely specified by the transition probabilities $\mathcal{P}_{j\beta}^{i\alpha}$, which account for the probabilities that a random walker at node $i$ in layer $\alpha$ moves to vertex $j$ in layer $\beta$. The master equation of this process reads[2]

[2] For convenience, we will not use the Einstein summation convention throughout Section 1.3.

$$p_{j\beta}(t+\Delta t) = \sum_{\alpha=1}^{L}\sum_{i=1}^{N}\mathcal{P}_{j\beta}^{i\alpha}p_{i\alpha}(t), \tag{1.11}$$

where $p_{i\alpha}(t)$ represents the probability of the random walker being in node $i$ of layer $\alpha$ at time $t$. Equation 1.11 expresses that the probability of being in $(j,\beta)$ at time $t+\Delta t$ is equal to the probability of being in any other $(i,\alpha)$ at time $t$, and then jumping to $(j,\beta)$. The double sum in Eq. 1.11 can be separated into four terms:

$$\begin{aligned} p_{j\beta}(t+\Delta t) = {} & \mathcal{P}_{j\beta}^{j\beta}p_{j\beta}(t) + \sum_{\substack{\alpha=1\\ \alpha\neq\beta}}^{L}\mathcal{P}_{j\beta}^{j\alpha}p_{j\alpha}(t) \\ & + \sum_{\substack{i=1\\ i\neq j}}^{N}\mathcal{P}_{j\beta}^{i\beta}p_{i\beta}(t) + \sum_{\substack{\alpha=1\\ \alpha\neq\beta}}^{L}\sum_{\substack{i=1\\ i\neq j}}^{N}\mathcal{P}_{j\beta}^{i\alpha}p_{i\alpha}(t), \end{aligned} \tag{1.12}$$

which take into account that, in the previous time step, the random walker could have already been in node $j$ and/or layer $\beta$.

Equation 1.11 can be put in a more compact form if we define $\mathbf{p}_\alpha$ as the row vector with $N$ components $p_{i\alpha}$ with respect to layer $\alpha$ and introduce the supra-vector $\mathbf{p} \equiv (\mathbf{p}_1, \mathbf{p}_2, \ldots, \mathbf{p}_L)$ with $NL$ components. Now, Eq. 1.11 can be written as

$$\dot{\mathbf{p}}(t) = -\mathbf{p}(t)\mathcal{L}, \tag{1.13}$$

hereafter referred to as the "random walk equation." In this equation, $\mathcal{L}$ refers to the $NL \times NL$ normalized supra-Laplacian matrix, whose structure is similar, although not identical, to the (unnormalized) supra-Laplacian matrix proposed in [119] to model the diffusion process in multiplex networks (see also [277]). The structure of the random walk equation is the same regardless of the transition probabilities $\mathcal{P}_{j\beta}^{i\alpha}$ adopted in Eq. 1.12. In particular, we are going to analyze four different prescriptions for the random walk dynamics: classical, diffusive, physical, and maximum entropy random walks.

### 1.3.2 Classical random walks

The classical description of random walkers on a graph (i.e., monoplex networks) is already present in [311, 290, 180], although applications to networks with complex topologies are more recent [214, 313].

In monoplex networks, the random walker has probability $1/k_i$ of moving from vertex $i$ to vertex $j$ in the neighborhood of $i$, where $k_i$ indicates the degree of a vertex $i$. The direct extension of such walks to the case of multiplex networks is considering the interlayer connections as additional edges available in vertex $i$. It follows that the probability of moving from vertex $i$ to vertex $j$ within the same layer $\alpha$ or of switching to the counterpart of vertex $i$ in layer $\beta$ is uniformly distributed. In such a scenario, the normalizing factor for obtaining the correct probability is the total strength $s_{i\alpha} + S_{i\alpha}$ of vertex $i$. The resulting transition probabilities for this classical random walker in a multiplex (RWC) are given in Table 1.1. For sake of completeness, the Laplacian matrix corresponding to this process in monoplex networks is generally referred to as the normalized Laplacian.

### 1.3.3 Diffusive random walks

In monoplex networks, this type of random walk has been studied in detail in [252]. Here, the random walker stays at vertex $i$ with a rate that depends on $i$. In fact, if $k_{\max} = \max_i\{k_i\}$ is the maximum degree in the network, the walker is allowed to wait in vertex $i$ with rate $1 - k_i/k_{\max}$ and to jump to any neighbor with rate $1/k_{\max}$. This procedure is equivalent to adding a weighted self-loop to each node in such a way that all nodes have the same strength. It can be shown that the corresponding random walk equation depends on the unnormalized Laplacian matrix, as in the classical diffusive process, hence the name "diffusive random walk."

We extend this walk to the case of multiplex networks by considering interlayer connections as additional edges to estimate the maximum vertex strength, $s_{\max} = \max_{i,\alpha}\{s_{i\alpha} + S_{i\alpha}\}$. The resulting transition rules for this random walker in a multiplex (RWD) are given in Table 1.1.

### 1.3.4 Physical random walks

Here, we propose a new type of random walk dynamics in the multiplex, which reduces to the classical random walk in the case of monoplex. The transition rules are the same,

**Table 1.1** *Transition probability for four different random walk processes on multiplex networks. We account for jumping between vertices (Latin letters) and switching between layers (Greek letters). When appearing in pairs, $j \neq i$ and $\beta \neq \alpha$ must be considered. See text for further detail.*

| Transition Probability | RWC | RWD | RWP | RWME |
|---|---|---|---|---|
| $\mathcal{P}_{i\alpha}^{i\alpha}$ | $\frac{D_{(i)}^{\alpha\alpha}}{s_{i\alpha} + S_{i\alpha}}$ | $\frac{s_{\max} + D_{(i)}^{\alpha\alpha} - s_{i\alpha} - S_{i\alpha}}{s_{\max}}$ | 0 | $\frac{D_{(i)}^{\alpha\alpha}}{\lambda_{\max}}$ |
| $\mathcal{P}_{i\beta}^{i\alpha}$ | $\frac{D_{(i)}^{\alpha\beta}}{s_{i\alpha} + S_{i\alpha}}$ | $\frac{D_{(i)}^{\alpha\beta}}{s_{\max}}$ | 0 | $\frac{D_{(i)}^{\alpha\beta}}{\lambda_{\max}} \frac{\psi_{(\beta-1)N+i}}{\psi_{(\alpha-1)N+i}}$ |
| $\mathcal{P}_{j\alpha}^{i\alpha}$ | $\frac{W_{ij}^{(\alpha)}}{s_{i\alpha} + S_{i\alpha}}$ | $\frac{W_{ij}^{(\alpha)}}{s_{\max}}$ | $\frac{W_{ij}^{(\alpha)}}{s_{i\alpha}} \frac{D_{(i)}^{\alpha\alpha}}{S_{i\alpha}}$ | $\frac{W_{ij}^{(\alpha)}}{\lambda_{\max}} \frac{\psi_{(\alpha-1)N+j}}{\psi_{(\alpha-1)N+i}}$ |
| $\mathcal{P}_{j\beta}^{i\alpha}$ | 0 | 0 | $\frac{W_{ij}^{(\beta)}}{s_{i\beta}} \frac{D_{(i)}^{\alpha\beta}}{S_{i\alpha}}$ | 0 |

*Abbreviations:* RWC, classical random walker; RWD, diffusive random walker; RWME, maximal entropy random walker; RWP, physical random walker.

except that we assume that the timescale for switching layers is negligible with respect to the timescale required to move from one vertex to another one in its neighborhood. Therefore, in the same time step, the random walker is allowed to both switch layers and jump to another vertex, with the layer-switching and vertex-jumping actions being independent. This is a fundamental difference from the random walkers described so far, because they were not allowed to both switch and jump in the same time unit. Moreover, another major difference lies in treating interlayer connections as another type of edge, one that does not compete with the intralayer edges.

As an example of this dynamics, one might imagine the case of online social networks where each layer corresponds to a different social structure (e.g., Facebook and Twitter), and users play the role of vertices. In this case, the time required by a user to switch from one layer to the other one requires less than a few seconds.

The resulting transition rules for this physical random walker in a multiplex (RWP) are given in Table 1.1. It is straightforward to show that this process is equivalent to the classical random walker in the case of monoplexes.

### 1.3.5 Maximal entropy random walks

In classical random walks, a walker jumps from a vertex to a neighbor with uniform probability that depends only on the local structure, namely, the vertex strength. However, a walk dynamics has recently been proposed where the transition rate of jumps is influenced by the global structure of the network [49], or only local information is available [267]. More specifically, the walkers choose the next vertex to jump into so as to maximize the entropy of their path at a global level, whereas classical random walkers maximize the entropy of their path at neighborhood level. To achieve such maximal entropy paths, the transition rates are governed by the largest eigenvalue of the adjacency matrix, and the components of the corresponding eigenvector [49].

In the case of multiplex networks, we use the supra-adjacency matrix Eq. 1.9 to achieve the same result. We indicate with $\lambda_{\text{max}}$ the largest eigenvalue of this matrix and with $\psi$ the corresponding eigenvector. Therefore, according to the prescription given in [49], the resulting transition rules for this maximal entropy random walker in a multiplex (RWME) are given in Table 1.1.

### 1.3.6 Comparison of types of random walkers

A representative example of each walk is shown in Figure 1.4, where vertices and layers visited by one random walker up to 100 time steps are reported. We show two different cases, corresponding to different choices of interlayer weights, to make evident the differences in the dynamics.

In the top panels of Figure 1.5, we show the transition probabilities in the case of a multiplex of 20 vertices embedded in two different realizations of a Watts–Strogatz small-world network [308]. The probability of finding a random walker in a certain vertex on a certain layer is also shown in the same figure, considering one walk starting from the first vertex only (middle panels) and from any other vertex with uniform probability (bottom

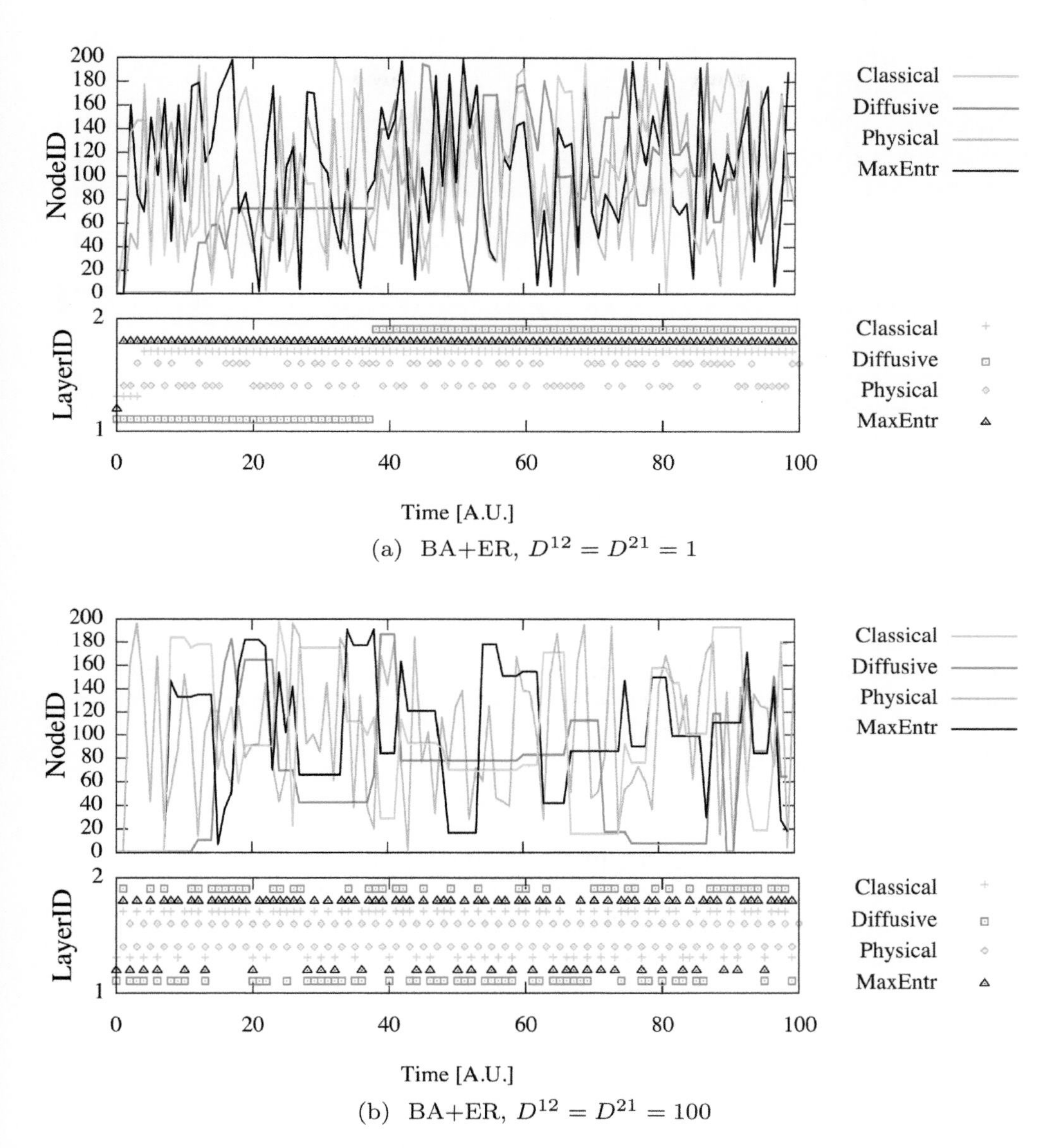


(a) BA+ER, $D^{12} = D^{21} = 1$



(b) BA+ER, $D^{12} = D^{21} = 100$

**Figure 1.4** *Random walks realizations on different multiplex structures. Vertices (top panels) and layers (bottom panels) visited by one random walker in 100 time steps. The four types of walk considered in this study are shown. The multiplex is built with one Barabási–Albert (BA, Layer 1) and one Erdős–Rényi (ER; Layer 2) network with 200 vertices, while interlayer weights are specified above.*

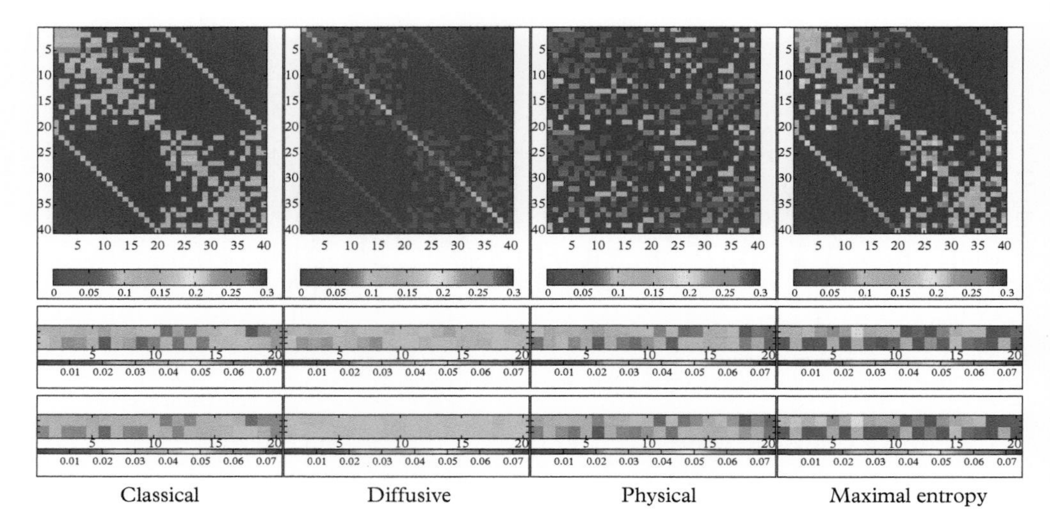


**Figure 1.5** *Probabilities governing four random walk strategies on multiplex networks. (Top) Transition probabilities for walks considered in this study. Note that we have rescaled by a factor of 2 the transition matrix of the diffusive walk for better visualization and to allow comparisons. (Middle) Occupation probability, for each vertex in each layer, considering one random walk starting only from the first vertex. (Bottom) As in the middle panels, but considering one random walk starting with uniform probability from any other vertex. Multiplex of 20 vertices embedded in two different realizations of a Watts–Strogatz small-world network (rewiring probability is 0.2), where* $D^{11} = D^{12} = D^{21} = D^{22} = 1$. *Different exploration strategies are responsible for the different probability that a vertex is visited and occupied by a random walker.*

panels). As expected, different exploration strategies result in different occupation probabilities, where some vertices in a certain layer might be explored more (or less) frequently, as in the case of RWC, RWP, and RWME, or uniformly, as in the case of RWD.

Figures 1.4 and 1.5 clearly highlight the different dynamics and how navigation strategy influences the exploration of the multiplex.

### 1.3.7 Occupation probability of random walkers

We define the occupation probability of node $i$ in layer $\alpha$ as the probability of finding the random walker in that location of the multiplex, in the limit of large time, $\Pi_{i\alpha} = \lim_{t\to\infty} p_{i\alpha}(t)$. We also indicate with $\Pi$ the corresponding supra-vector. In general, $\Pi$ is the left eigenvector of the supra-transition matrix corresponding to the unit eigenvalue. In some cases, the occupation probability can be estimated from the detailed balance equation

$$\Pi_{i\alpha}\mathcal{P}^{i\alpha}_{j\beta} = \Pi_{j\beta}\mathcal{P}^{j\beta}_{i\alpha}, \tag{1.14}$$

obtaining

$$\Pi_{i\alpha} = \frac{s_{i\alpha} + S_{i\alpha}}{\sum_\beta \sum_j (s_{j\beta} + S_{j\beta})} \tag{1.15}$$

for RWC, generalizing the well-known result obtained for walks in a monoplex network,

$$\Pi_{i\alpha} = \frac{1}{NL} \tag{1.16}$$

for RWD, as expected for a purely diffusive walk, and

$$\Pi_{i\alpha} = \psi^2_{(\alpha-1)N+i} \tag{1.17}$$

for RWME, generalizing the results obtained in [49] for monoplex networks.

Indeed, following the approach proposed in [214] for random walks on monoplexes, it is possible to show that the time required for a random walker starting from vertex $i$ to arrive back to the same vertex, that is, the mean return time, is given by

$$\langle T_{ii} \rangle = \frac{1}{\sum\limits_{\alpha=1}^{L} \Pi_{i\alpha}}. \tag{1.18}$$

It is straightforward to verify that distributions expected in the case of monoplex are recovered for $L = 1$. It is worth noting that for classical random walks, the occupation probability of vertex $i$ is proportional to its supra-strength, that is, intra- plus interlayer strengths, whereas for diffusive walks, such a probability is the same for any vertex, regardless of multiplex topology.

### 1.3.8 Random walks coverage

The coverage of a random walk is an important measure which quantifies how difficult (or easy) is to visit all the nodes in a network. There are two main approaches for defining the coverage: (i) calculate the expected time the random walker takes to visit all the nodes; (ii) calculate the average fraction of visited nodes as a function of the length of the walk. Here, we adopt the second option, which is computationally friendlier, and denote the coverage as $\rho(t)$. In multiplex networks, one has to take into account that the same node is present in all layers; thus, the logical approach is to consider a node to have been "visited" if the random walker visited that node at least once in any of the layers. For example, in a multiplex transportation network consisting of buses, trains, and a metro, it is not important if you have arrived at a location by bus or metro; what counts is having been there.

First, the probability of finding the random walker at node $i$ at time $t$, regardless of the layer, is given by $p_i(t) = \sum_\alpha p_{i\alpha}(t)$. Introducing the supra-vector $\mathbf{E}_i \equiv (\mathbf{e}_i, \mathbf{e}_i, \ldots, \mathbf{e}_i)$, where $\mathbf{e}_i$ is the $i$th canonical vector, we may write

$$p_i(t+1) = \mathbf{p}(t)\mathcal{P}\mathbf{E}_i^{\dagger}. \tag{1.19}$$

We have set $\Delta t = 1$ to make $t$ equivalent to the walk length. Next, we define the probability $\sigma_{ij}(t)$ of not finding the random walker at node $i$ after $t$ time steps, assuming that it started at vertex $j$. This probability satisfies the recursive relation

$$\sigma_{ij}(t+1) = \sigma_{ij}(t)\,[1 - p_i(t+1)], \tag{1.20}$$

which can be written as

$$\dot{\sigma}_{ij}(t) = -\sigma_{ij}(t)\mathbf{p}(t)\mathcal{P}\mathbf{E}_i^{\dagger} \tag{1.21}$$

and whose solution is

$$\sigma_{ij}(t) = \sigma_{ij}(0)\exp\left[-\mathbf{p}_j(0)\mathbb{P}\mathbf{E}_i^{\dagger}\right], \quad \mathbb{P} = \sum_{\tau=0}^{t} \mathcal{P}^{\tau+1}, \tag{1.22}$$

where $\mathbf{p}_j(0) \equiv (\mathbf{e}_j, \mathbf{0}, \ldots, \mathbf{0})$ explicitly indicates that at time $t = 0$, the walker started at vertex $j$ in the first layer, without loss of generality. The matrix $\mathbb{P}$ accounts for the probability of reaching each vertex through any path of length $1, 2, \ldots, t+1$. Note also that $\sigma_{ij}(0) = 1 - \delta_{ij}$, where $\delta_{ij}$ is the Kronecker delta, since a walk starting at node $i$ cannot also at the same time be at node $j$ unless $i = j$. Finally, a good approximation of the coverage is given by double averaging over all vertices the probability $1 - \sigma_{ij}(t)$, obtaining

$$\rho(t) = 1 - \frac{1}{N^2}\sum_{\substack{i,j=1 \\ i \neq j}}^{N} \exp\left[-\mathbf{p}_j(0)\mathbb{P}\mathbf{E}_i^{\dagger}\right], \tag{1.23}$$

which can be solved numerically to obtain the coverage at each time step. Comparisons between the predicted coverage using Eq. 1.23 and Monte Carlo simulations show a perfect agreement, thus confirming the validity of our theoretical development (see [73]). It must be pointed out that this expression for the coverage is applicable not only to multiplex networks but also to standard single-layer networks.

We show in Figure 1.6 the coverage versus time in the case of RWP only, for some representative multiplexes where $D_{(i)}^{12} = D_{(i)}^{21} = D_{(i)}^{11} = D_{(i)}^{22} = 1, \forall i = 1, \ldots, N$. Results for different combination of topologies (indicated by double acronyms in the figure) are shown, together with results for walks in a single layer (indicated by single acronyms in the figure). The topologies analyzed include Erdős–Rényi (ER), Barabási–Albert (BA), and Watts–Strogatz (WS) networks. The abbreviation "diff" indicates that the layers have the same topology but different random realizations, while "same" indicates that the same topology and same random realization is present in both layers. The inset shows the relative difference of coverages with respect to the case of an ER monoplex.

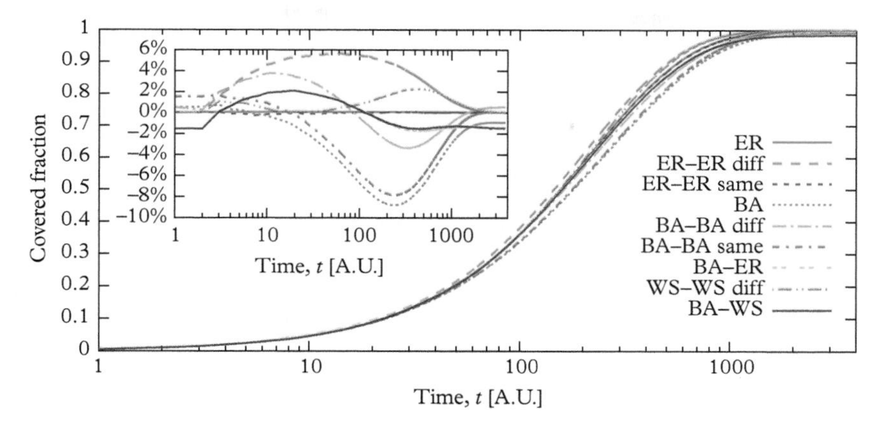


**Figure 1.6** *Dependence of the coverage on multiplex topology: number of visited vertices versus time for monoplex and multiplex topologies (see the text for further details). The inset shows the relative difference of each curve with respect to the coverage obtained for an ER monoplex, showing that vertices in different topologies are visited with different timescales.*

The multiplex topology has an evident impact on the walk process, delaying or accelerating the exploration of the network with respect to a random search in a monoplex random network. This is a genuine effect of the multilayer structure, and it is not related to the finite size of the considered networks, as shown in Figure 1.7, where multiplexes of 2,000 nodes and many different topologies are considered.

In Figure 1.8, for each random walk considered, we show the inverse of the time $\tau_C$ required to cover 50% of a BA+ER multiplex with 200 vertices as a function of the interlayer weight $D_X = D^{12} = D^{21}$, with $D^{11} = D^{22}$. It is worth mentioning that the final result depends quantitatively, but not qualitatively, on the choice of the covered fraction. This representative example shows the impact of transition rules on the exploration of the multiplex, providing evidence that the best strategy to use to cover the network depends on the topology and on the weight of inter-layer connections. Moreover, in this specific experiment, the walk in the multiplex is *infra-diffusive* (*sub-diffusive*), depending on the value of $D_X$, that is, the time to cover the multiplex lies between (is smaller than) the times required to cover each layer separately. It is worth noting that in other cases, like RWME on BA+BA multiplexes, walks show *enhanced diffusion*, that is, the time to cover the multiplex is smaller than the time to cover each layer separately. This is shown, for instance, in Figure 1.9.

## 1.4 Centrality and versatility

In this section, we focus on the definition of node centrality in multilayer networks. We obtain these properties using algebraic operations involving the multilayer adjacency tensor, canonical vectors, and canonical tensors, achieving the natural extension of the

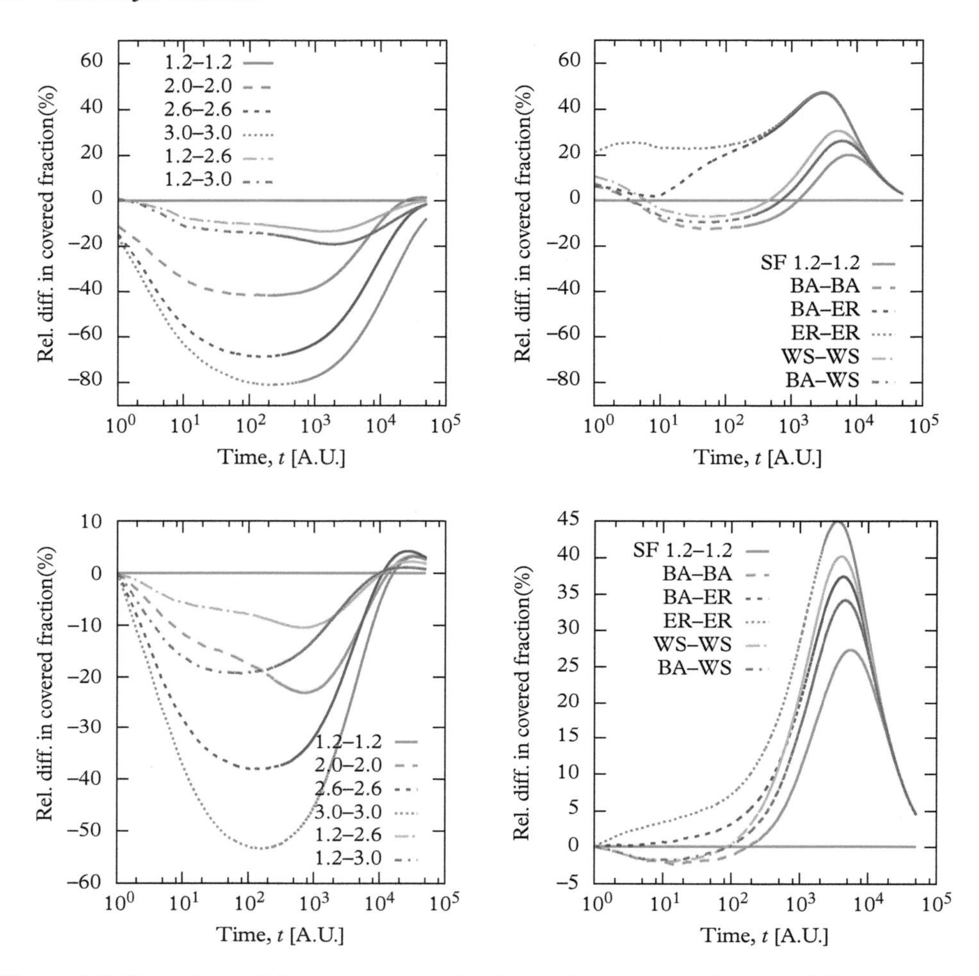


**Figure 1.7** *Dependence of the coverage on multiplex topology: same as the inset of Figure 1.6, where the relative difference of each curve is calculated with respect to the coverage obtained for a multiplex of two different scale-free (SF) networks with exponent 1.2. Top panels refer to RWC, whereas bottom panels refer to RWP. Left panels (top and bottom) refer to multiplexes of different scale-free networks with other degree distributions, whose exponents are specified in the legend. Right panels (top and bottom) refer to multiplexes of other topologies.*

concept of centrality in single-layer networks. We refer the reader to Ref. [72] for other multilayer network diagnostics.

In practical applications, one is often interested in assigning a global measure of importance to each node, aggregating the information obtained from the different layers. A naive choice could be to combine the centrality of the nodes—obtained from the different layers separately—according to some heuristic choice. This is a viable

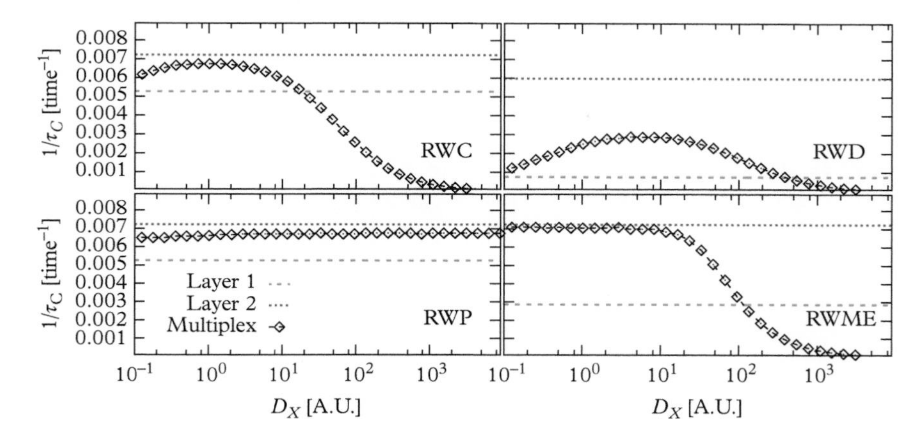


**Figure 1.8** *Critical dependence of the coverage on navigation strategy and interlayer connection strength. Different random walks are used to calculate the inverse of the time $\tau_C$ required to cover 50% of a BA+ER multiplex with 200 vertices, as a function of $D_X = D^{12} = D^{21}$. The values for walks in each layer are shown for comparison and make clear how different exploration strategies have a strong effect on the coverage timescale.*

solution when there is no interconnection between layers, that is, in the case of edge-colored graphs [274, 125]. However, the main drawback of applying this approach to interconnected multilayer networks is that the measure will depend on the choice of the heuristics and might not evaluate the real importance of nodes. Conversely, our approach capitalizes on the tensorial formulation of interconnected multilayer networks and accounts for the higher level of complexity of such systems without relying on external assumptions and naturally extending the well-known centrality measures adopted for several decades in the case of monoplexes.

## 1.4.1 Eigenvector centrality

Among the numerous notions of centrality introduced to quantify the importance of nodes (and other components) in a network [307], eigenvector centrality is among the oldest ones [43, 42]. The eigenvector centrality of a node is defined to be proportional to the sum of the eigenvector centralities of its neighbors. The recursive nature of this notion yields a vector of centralities that satisfies an eigenvalue problem. In the case of monoplexes, the eigenvector centrality vector $\mathbf{v}$, whose components are the centralities of nodes according to [43, 42], is a solution of the eigenvector equation $W^{i}_{j} v_i = \lambda_1 v_j$, where $\lambda_1$ is the largest eigenvalue of $W^{i}_{j}$, and $v_i$ indicates the eigenvector centrality of node $i$ (note the use of the Einstein summation convention).

A naive approach for the calculation of the importance of each node in an interconnected network might be to project the interconnected topology to an aggregated monoplex and to associate to each node the centrality that node has in such an aggregated

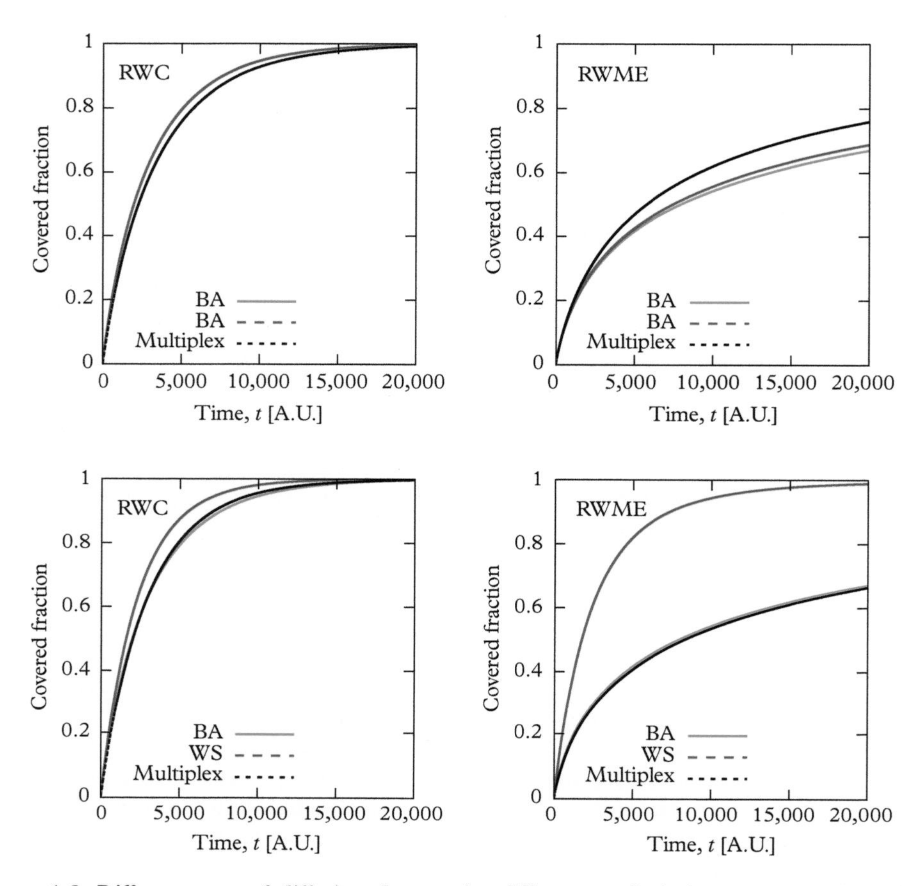


**Figure 1.9** *Different types of diffusion characterize different topological structures and navigation strategies. Coverage versus time for two different multiplex topologies (BA+BA on the top panels, and BA+WS on the bottom panels) and two different walk rules (RWC on the left panels, and RWME on the right panels). While the diffusion on single layers separately and that on the multiplex are similar for RWC on BA+BA, this is not the case for RWME on BA+BA, where enhanced diffusion is shown in the multiplex. In the other cases, the diffusion is* infra-*diffusive.*

network. The main drawback of this approach is that it mixes the information from all layers, with uncontrollable effects, as shown in Ref. [74] for both synthetic and empirical networks. Another attempt to extend this calculation to the case of multilayer networks might be to calculate the eigenvector centralities for each layer separately, to build the tensor $\bar{V}_{i\alpha}$ encoding the centrality of each node in each layer. The final step would be to choose a heuristic aggregation of such centralities to assign a unique centrality measure to each node, regardless of the layer. However, the tensor $\bar{V}_{i\alpha}$ is not the solution of a unique eigenvalue problem but the combination of the solutions of $L$ different eigenvalue problems treated separately; therefore, it is not a natural extension of the notion of eigenvector centrality to the realm of interconnected multilayer networks.

Instead, according to Ref. [72], this descriptor can be obtained as the solution of the tensorial equation

$$M_{j\beta}^{i\alpha}\Theta_{i\alpha} = \lambda_1 \Theta_{j\beta}, \tag{1.24}$$

where $\lambda_1$ is the largest eigenvalue, and $\Theta_{i\alpha}$ is the corresponding *eigentensor* encoding the centrality of each node in each layer when accounting for the whole interconnected structure. The eigentensor can be obtained by means of an iterative procedure, as the power method in the case of monoplexes. An analysis of this eigentensor problem is provided in Section 1.2.3. Thus, the multilayer generalization of Bonacich's eigenvector centrality [43, 42] is given by Eq. 1.24.

This centrality, like others in the rest of this section, assigns a measure of importance to each node in each layer, accounting for the full interconnected structure of the multilayer network. However, in practical applications, one is often interested in assigning a global measure of importance to each node, aggregating the information obtained from the different layers. The choice of the aggregation method is not trivial; it strongly influences the final estimation and might lead to wrong results. However, this is not case for the tensorial framework discussed so far. In fact, the centrality $\Theta_{i\alpha}$ is calculated by inherently accounting for the interconnected structure of the whole system. We do not need to arbitrarily combine the information from different separate measures. In our framework, the most intuitive type of aggregation, that is, summing up over layers, represents the unique and correct choice. Thus, the eigenvector centrality of each node becomes $\theta_i = \Theta_{i\alpha}u^{\alpha}$, where $u^{\alpha}$ is the tensor with all components equal to 1.

### 1.4.2 Katz centrality

It is a well-known fact that eigenvector centrality can lead to incorrect results in the case of directed networks. In fact, nodes with only outgoing edges have an eigenvector centrality of 0 if Bonacich's definition is adopted. Moreover, in this case, there are two leading eigenvectors, for incoming centrality and outgoing centrality, requiring distinguishing between covariant and contravariant calculations. The Katz centrality [144] solves the above problem by assigning a small amount $b$ of centrality to each node before calculating centrality. For monoplexes, the Katz centrality is given by $v_j = aW_j^i v_i + bu_j$, where $a$ must be smaller than the reciprocal of the largest eigenvalue $\lambda_1$ of $W$, and one often chooses $b = 1$.

Following a similar idea, we define the centrality tensor for each node in each layer as the solution of the tensorial equation

$$\Phi_{j\beta} = aM_{j\beta}^{i\alpha}\Phi_{i\alpha} + bu_{j\beta}, \tag{1.25}$$

corresponding to the natural extension of the equation proposed by Katz to the case of interconnected multilayer networks. The solution is given by $\Phi_{j\beta} = [(\delta - aM)^{-1}]_{j\beta}^{i\alpha} U_{i\alpha}$, where $\delta_{j\beta}^{i\alpha} = \delta_j^i \delta_\beta^\alpha$. As for the eigentensor centrality, this *Katz centrality tensor* accounts for the whole interconnected topology and it is enough to contract it with the 1-vector to obtain the Katz centrality for each node, that is, $\phi_i = \Phi_{i\alpha} u^{\alpha}$.

### 1.4.3 Hubs and authorities centrality

In directed networks, such as Web sites, we can rank nodes differently according to their importance as senders or receptors of links, respectively. The Hyperlink-Induced Topic Search (HITS) approach, also known as the hubs and authorities' algorithm [150], assigns two different descriptors for each node, namely, hub and authority. In fact, Web pages that point to an important page generally also point to other important pages, building a structure similar to a bipartite topology where relevant pages—that is, authorities—are pointed by special Web pages—that is, hubs. It follows that nodes with high authority centrality are linked by nodes with high hub centrality, while very influent hubs point to nodes that are very authoritative. Such a mechanism is described by two coupled equations which reduce to the two eigenvalue problems $(WW^{\dagger})^i_j v_i = \lambda_1 v_j$ and $(W^{\dagger} W)^i_j z_i = \lambda_1 z_j$, where $W^{\dagger}$ denotes the transpose of the adjacency matrix, $\lambda_1$ indicates the leading eigenvalue, and $v_i$ and $z_i$ indicate hub and authority scores, respectively. The natural extension of the equations proposed by Kleinberg to the case of interconnected multilayer networks is given by

$$(MM^{\dagger})^{i\alpha}_{j\beta} \Gamma_{i\alpha} = \lambda_1 \Gamma_{j\beta}, \tag{1.26}$$

$$(M^{\dagger}M)^{i\alpha}_{j\beta} \Upsilon_{i\alpha} = \lambda_1 \Upsilon_{j\beta}, \tag{1.27}$$

where $\Gamma_{i\alpha}$ and $\Upsilon_{i\alpha}$ indicate hub and authority centrality, respectively. It is worth remarking that for undirected interconnected multiplexes, hub and authority scores are the same as and equal to the corresponding eigenvector centrality. The hub and authority tensors should be contracted with the 1-vector to obtain the scores corresponding to each node regardless of the layer, that is, $\gamma_i = \Gamma_{i\alpha} u^{\alpha}$, and $\upsilon_i = \Upsilon_{i\alpha} u^{\alpha}$, respectively.

### 1.4.4 Random walk centralities

Random walks can also be used to calculate the centrality of actors in complex networks, for example, when there is no knowledge of the full topology, and only local information is available. In such cases, centrality descriptors based on the shortest paths, for example, betweenness and closeness centrality, should be substituted by centrality notions based on random walks [214, 211].

As we have seen in Section 1.3, a random walk on a multilayer network induces nontrivial effects because the presence of interlayer connections affects its navigation of a networked system [73]. Let $\mathcal{P}^{i\alpha}_{j\beta}$ denote the tensor of transition probabilities for jumping between pairs of nodes and switching between pairs of layers, and let $p_{i\alpha}(t)$ be the time-dependent tensor that gives the probability of finding a walker at a particular node in a particular layer. Hence, the covariant master equation that governs the discrete-time evolution of the probability from time $t$ to time $t+1$ is given by Eq. 1.11, which reads $p_{j\beta}(t+1) = \mathcal{P}^{i\alpha}_{j\beta} p_{i\alpha}(t)$.

The steady-state solution of this equation is given by $\Pi_{i\alpha}$, quantifying the probability of finding a walker in the node $i$ of layer $\alpha$ in the infinite-time limit. In the case of monoplexes, the steady-state solution can be obtained by solving the eigenvalue problem for the rank-2 transition tensor and calculating the leading eigenvector corresponding to the unitary eigenvalue. Similarly, in the case of multilayer networks, the solution can be obtained by calculating the leading *eigentensor*, solution of the higher-order eigenvalue problem

$$\mathcal{P}^{i\alpha}_{j\beta} \Pi_{i\alpha} = \lambda \Pi_{j\beta}. \tag{1.28}$$

We refer to Section 1.2.3 for the mathematical details to solve this problem. The probability $\Pi_{i\alpha}$, which we define as *random walk occupation centrality*, depends on the full interconnected structure of the multilayer network and, likewise, the previously described multilayer centralities. Finally, we may aggregate by layer to obtain the corresponding node centralities, $\pi_i = \Pi_{i\alpha} u^{\alpha}$.

Although different exploration strategies can be adopted to walk in a multilayer network [73], we first focus on the classical random walks (RWC) as previously described in Section 1.3.2. Let us indicate with $\Omega_{i\alpha}$ the strength of node $i$ in layer $\alpha$, including the interlayer connections, that is, $\Omega_{i\alpha} = s_{i\alpha} + S_{i\alpha}$, where $s_{i\alpha}$ and $S_{i\alpha}$ are the intralayer and interlayer strengths, respectively. The multi-strength vector, whose components indicate the strength of each node accounting for the full multilayer structure, is given by summing up its strengths across all layers, that is, by $\omega_i = \Omega_{i\alpha} u^{\alpha}$. We indicate with $D^{i\alpha}_{j\beta}$ the strength tensor whose entries are all zeros, except for the $i = j$ and $\alpha = \beta$ entries, which are given by $\Omega_{i\alpha}$. This tensor represents the multilayer extension of the well-known diagonal strength matrix in the case of monoplexes. Therefore, the transition tensor is given by $\mathcal{P}^{i\alpha}_{j\beta} = M^{k\gamma}_{j\beta} \tilde{D}^{i\alpha}_{k\gamma}$, where $\tilde{D}^{i\alpha}_{j\beta}$ is the tensor whose entries are the reciprocals[3] of the non-zero entries of the strength tensor. For this classical random walk, it can be easily shown that $\Pi_{i\alpha} \propto \Omega_{i\alpha}$ [73].

It is worth noting that, in this specific case, the computation of the centrality by means of the aggregated network would provide the same centralities for the interconnected multiplex, if interlayer edges were replaced by self-loops. In the more specific case that the interlayer edges have the same strength for all nodes, the random walk centrality will be just a linear function of the strengths in the aggregated network, without the necessity of accounting for the self-loops, thus recovering the traditional *degree centrality* [265, 101] for unweighted, undirected monoplex networks. However, this is no longer the case for the other centrality measures discussed in this section, where calculating the diagnostics from the aggregate might lead to wrong conclusions.

[3] It is worth remarking that, in general, this is different from the inverse of a tensor $A^{i\alpha}_{j\beta}$, which is defined as the tensor $B^{i\alpha}_{j\beta}$ such that $A^{i\alpha}_{k\gamma} B^{k\gamma}_{j\beta} = \delta^{i\alpha}_{j\beta}$, where $\delta^{i\alpha}_{j\beta} = \delta^{i}_{j} \delta^{\alpha}_{\beta}$.

For the other types of random walkers, namely diffusive (RWD), physical (RWP), and maximal entropy (RWME), see Section 1.3 and, in particular, Section 1.3.7, which provide the different occupation probabilities, leading to the respective alternative definitions of random walk occupation probability centralities.

Apart from centralities derived from random walk occupation probabilities, it is possible to define additional centralities based on other properties of the random walkers. The most relevant are *PageRank* [46], which will be described in Section 1.4.5, *random walk betweenness centrality* and *random walk closeness centrality* [211]. Random walk betweenness measures the net flow of random walkers through nodes, and random walk closeness is related to the mean first passage time needed to reach a node from the rest of locations in the network. See Ref. [276] for their extension to multilayer interconnected networks.

### 1.4.5 PageRank centrality

We capitalize on the previous analysis of random walkers to extend to interconnected networks a widely adopted measure of centrality, that is, the *PageRank centrality* [46]. A recent study in this direction has been reported in [125], in the case of edge-colored graphs, where the authors, exploiting the random walk interpretation of PageRank centrality, define the PageRank of a multiplex network by means of a random walk subject to teleportation. In that study, the PageRank for nodes in the first layer is computed using the standard definition for a monoplex [46], whereas the PageRank for nodes in the second layer is computed using the centrality information obtained from the first one. It is worth noting that this definition is limited to edge-colored graphs with only two layers, with any extension to a larger number of layers being possible but very complicated from the mathematical point of view.

Here, we exploit the fact that PageRank centrality can be seen as the steady-state solution of the equation $p_j(t+1) = R^i_j p_i(t)$ in the case of monoplexes, where $R^i_j$ is the transition matrix of a random walk where the walker jumps to a neighbor with rate $r$ and then teleports to any other node in the network with another rate, $1-r$. In the case of interconnected multilayer networks, the teleportation might occur to any other node in any layer. Depending on the application of interest, the walker can be teleported to other nodes with a rate that is specific to each layer. However, to keep the study as simple as possible, we consider the case with the same teleportation rate for all layers. Let $\mathcal{R}^{i\alpha}_{j\beta}$ be the corresponding transition tensor, where the walker jumps to a neighbor with rate $r$ and teleports to any other node in the network with the rate $1-r$. This rank-4 tensor is given by

$$\mathcal{R}^{i\alpha}_{j\beta} = r\mathcal{P}^{i\alpha}_{j\beta} + \frac{1-r}{NL} u^{i\alpha}_{j\beta}, \tag{1.29}$$

where $u^{i\alpha}_{j\beta}$ is the rank-4 tensor with all components equal to 1. The steady-state solution of the master equation corresponding to this transition tensor provides the PageRank centrality for interconnected multiplex networks. This definition is valid for all multiplexes where all nodes have outgoing edges. If this is not the case, as in several real-world networks, Eq. (1.29) reduces to $\mathcal{R}^{i\alpha}_{j\beta} = \frac{1}{NL}u^{i\alpha}_{j\beta}$ for all nodes $i$ with no outgoing connections, ensuring the correct normalization of the transition tensor $\mathcal{R}^{i\alpha}_{j\beta}$. We set $r = 0.85$, as in the classical PageRank algorithm.

To compute the aggregate centrality of a node, accounting for the whole interconnected topology, we proceed as for the random walk occupation centrality previously discussed. Let $\Pi_{i\alpha}$ be the eigentensor of the transition tensor $\mathcal{R}^{i\alpha}_{j\beta}$ (see Section 1.2.3 for details), denoting the steady-state probability of finding the walker in node $i$ and layer $\alpha$. The multilayer PageRank is obtained by simply contracting the layer index of the eigentensor with the 1-vector: $\pi_i = \Pi_{i\alpha}u^{\alpha}$, that is, by summing up over layers.

### 1.4.6 Centrality measures based on shortest path

For the sake of completeness, we briefly consider here centrality measures based on shortest paths, namely, betweenness and closeness.

The extension of the *shortest-path betweenness centrality*, defined in the case of monoplex networks in Refs [12, 100, 45], is obtained by counting the number of shortest paths between any pair of *origin* and *destination* nodes $(o, d)$ that go through node $j$ in the interconnected structure [275, 73].

Equivalently to the case of a monoplex, we define a path $\ell_{[o\sigma\to d\gamma]} \in \mathcal{P}_{[o\sigma\to d\gamma]}$, in the interconnected multilayer network, as an ordered sequence of nodes which starts from node $o$ in layer $\sigma$ and finishes in node $d$ in layer $\gamma$. We require that there exists an edge between all pairs of consecutive nodes in $\ell$. Here, $\mathcal{P}_{[o\sigma\to d\gamma]}$ indicates the set of all possible paths between node $o$ in layer $\sigma$ and node $d$ in layer $\gamma$. For every path $\ell_{[o\sigma\to d\gamma]}$, it is possible to define a cost function $c(\ell_{[o\sigma\to d\gamma]})$, usually depending on the weight of the edges the path traverses and on the application of interest, to account for the "goodness" of the path. Hence, the shortest path from node $o$ in layer $\sigma$ to node $d$ in layer $\gamma$ is the path

$$\ell^*_{[o\sigma\to d\gamma]} = \min_{\ell'_{[o\sigma\to d\gamma]}\in\mathcal{P}_{[o\sigma\to d\gamma]}} c(\ell'_{[o\sigma\to d\gamma]}), \tag{1.30}$$

which minimizes the cost function. Using Eq. (1.30), we define the shortest path from node $o$ to node $d$, regardless of the layer, as

$$\ell^*_{[o\to d]} = \min_{\sigma,\gamma\in\{1,\ldots,L\}} \ell^*_{[o\sigma\to d\gamma]}. \tag{1.31}$$

The shortest-path betweenness centrality $\tau_j$ of node $j$ is defined to be proportional to the number of times that node $j$, regardless of the layer, belongs to the set $\ell^*_{[o\to d]}$ for every possible origin–destination pair $(o, d)$. We must remark that betweenness centrality is crucial for understanding congestion in networks [278].

On the other hand, in the same spirit of monoplex networks, we define the *shortest-path closeness centrality* (see [28, 249]) of a node $j$ in an interconnected multilayer topology as the average of the inverse of the cost of the shortest paths that start from any other node $o$

in the network [73]. Thus, given the cost of a shortest path $c(\ell^*_{[o\rightarrow i]})$ between node $i$ and node $o$, the shortest-path closeness centrality $\xi_i$ can be easily computed by considering all possible origin nodes $o$.

Note that the shortest paths contributing to betweenness and closeness centralities may start and/or end in only a few of the available layers, and that they may contain interlayer edges. This means that, once again, it is impossible to derive the correct centralities by just considering the aggregated network or the individual layers of the multilayer network.

### 1.4.7 Centrality becomes versatility

The calculation of centrality in several empirical multilayer interconnected networks shows that the highly ranked nodes are not those with large importance in the aggregated network or in individual layers, but the nodes responsible for the cohesion of the whole structure, bridging together different types of relations [73]. Thus, they can be called *versatile nodes*, and we can safely say that centrality becomes a measure of *versatility* for this kind of networks.

Table 1.2 contains the 25 nodes with largest PageRank versatility of a Wikipedia multilayer interconnected network consisting of biologists, chemists, computer scientists, economists, inventors, mathematicians, philosophers, and physicists [73]. The multilayer network is the largest connected component, formed by 5,513 nodes and 8 layers. Weighted links are established according to the hyperlinks found in the corresponding Web pages and can be either intralayer or interlayer links (see [73] for all the details).

The top-rated scientist is Edmund F. Robertson, due to his being one of the creators of the MacTutor History of Mathematics Archive, a Web site containing biographies of many mathematicians, whose corresponding pages point to this Web site and the Wikipedia page of Edmund F. Robertson. Thus, this node could be considered spurious, one which should have been removed from the network during the preprocessing of the data. Anyway, this is not important for assessing the meaning of versatility. For example, Milton Friedman made contributions to economics, statistics, international finance, risk/insurance, and microeconomic theory: Hilary Putnam is a computer scientist and mathematician with outstanding contributions in the philosophy of mind, of mathematics and of science; E. O. Wilson is the father of sociobiology; Harold Clayton Urey won the Nobel Prize in Chemistry and is well known for theories on the development of organic life from nonliving matter and for playing a significant role in the development of the atomic bomb; and Kurt Gödel is one of the greatest logicians of all time, with impacts on several different disciplines, from pure mathematics to physics and philosophy. Out of this distinguished group, Wilson and Clayton are ranked among the highest (with scores ranging from 300 to almost 1,000) in the aggregated network, and with respect to the average centrality of the separated layers. In contrast, despite being still highly rated, Albert Einstein and Plato are not as highly ranked when the full structure of the network is considered.

In summary, we have seen how taking into account the full structure of multilayer interconnected networks has important consequences for their structural properties and

**Table 1.2** *Top ranked nodes of Wikipedia dataset by PageRank versatility. For comparison purposes, the table also shows the corresponding ranks according to the PageRank in the aggregated network, and the ranks after averaging the PageRank at every single layer as independent networks. In parentheses, the variation of rank with respect to versatility is shown. Global diversity stands for the number of layers sending or receiving links to the considered node.*

| | PageRank Centrality Ranking | | | Diversity |
|---|---|---|---|---|
| **Name** | **Versatility** | **Aggregate** | **Average** | **Global** |
| Edmund F. Robertson | 1 | 1 (+0) | 18 (+17) | 2 |
| Milton Friedman | 2 | 16 (+14) | 22 (+20) | 8 |
| Hilary Putnam | 3 | 34 (+31) | 1,302 (+1,299) | 4 |
| E. O. Wilson | 4 | 332 (+328) | 996 (+992) | 8 |
| Harold Clayton Urey | 5 | 537 (+532) | 451 (+446) | 8 |
| Kurt Gödel | 6 | 43 (+37) | 325 (+319) | 8 |
| Avicenna | 7 | 30 (+23) | 8 (+1) | 4 |
| Ernst Mayr | 8 | 191 (+183) | 582 (+574) | 8 |
| Herbert A. Simon | 9 | 48 (+39) | 14 (+5) | 8 |
| Charles Stark Draper | 10 | 1,196 (+1,186) | 1,169 (+1,159) | 8 |
| Ivan Pavlov | 11 | 423 (+412) | 56 (+45) | 6 |
| Aristotle | 12 | 3 (−9) | 2 (−10) | 5 |
| Paul Samuelson | 13 | 26 (+13) | 43 (+30) | 8 |
| Immanuel Kant | 14 | 2 (−12) | 19 (+5) | 2 |
| Norbert Wiener | 15 | 68 (+53) | 407 (+392) | 8 |
| Chien-Shiung Wu | 16 | 100 (+84) | 599 (+583) | 6 |
| George Dantzig | 17 | 217 (+200) | 1,244 (+1,227) | 8 |
| Ronald Ross | 18 | 1,602 (+1,584) | 5,472 (+5,454) | 5 |
| John C. Slater | 19 | 1,242 (+1,223) | 1,627 (+1,608) | 8 |
| Porphyry (philosopher) | 20 | 311 (+291) | 1,063 (+1,043) | 3 |
| Peter Mansfield | 21 | 1,502 (+1,481) | 2,914 (+2,893) | 5 |
| Rosalyn Yalow | 22 | 888 (+866) | 1,580 (+1,558) | 7 |
| Samuel Goudsmit | 23 | 1,364 (+1,341) | 1,960 (+1,937) | 8 |
| Albert Einstein | 24 | 10 (−14) | 11 (−13) | 4 |
| Plato | 25 | 4 (−21) | 15 (−10) | 2 |

the dynamics from them, and these have boosted the interest in this kind of system and yielded an enormous amount of scientific literature.

## ACKNOWLEDGMENTS

We acknowledge funding from the European Commission FET-Proactive projects PLEXMATH (grant 317614) and MULTIPLEX (grant 317532), and the Spanish Ministerio de Economía y Competitividad (grant number FIS2015-71582-C2-1). A. A. also acknowledges financial support from the Generalitat de Catalunya ICREA Academia and the James S. McDonnell Foundation.

# 2

# Reconstructing Random Jigsaws

**Paul Balister[1], Béla Bollobás[1,2,3], and Bhargav Narayanan[2]**

[1]Department of Mathematical Sciences, University of Memphis, Memphis TN 38152, USA

[2]Department of Pure Mathematics and Mathematical Statistics, University of Cambridge, Wilberforce Road, Cambridge CB3 0WB, UK

[3]London Institute for Mathematical Sciences, 35a South St., Mayfair, London W1K 2XF, UK

## 2.1 Introduction

The reconstruction problem for a family of discrete structures asks the following: is it possible to uniquely reconstruct a structure in this family from the "deck" of all its substructures of some fixed size? Combinatorial reconstruction problems have a very rich history. The oldest such problem is perhaps the graph reconstruction conjecture of Kelly and Ulam [145, 299, 126], and analogous questions for various other families of discrete structures have since been studied; see, for instance, the results of Alon, Caro, Krasikov, and Roddity [8] on reconstructing finite sets satisfying symmetry conditions, Pebody's [227, 228] results on reconstructing finite abelian groups, and the results of Pebody, Radcliffe, and Scott [229] on reconstructing finite subsets of the plane.

Another natural line of enquiry, and the one we pursue here, is to ask how the answer to the reconstruction problem changes when we are required to reconstruct a *typical* (as opposed to an arbitrary) structure in a family of discrete structures. These probabilistic questions typically have substantially different answers, compared to their extremal counterparts, as evidenced by the results of Bollobás [40] and Radcliffe and Scott [235], for example.

Here, we shall study a reconstruction problem proposed by Mossel and Ross in connection with the problem of shotgun sequencing DNA. To state this problem, we need a few definitions. For $n \in \mathbb{N}$, we write $[n]$ for the set $\{1, 2, \ldots, n\}$, and by the *extended* $n \times n$ grid, we mean the grid $[n]^2 \subset \mathbb{Z}^2$ together with the edges of $\mathbb{Z}^2$ incident

Balister, P., Bollobás, B., and Narayanan, B., "Reconstructing Random Jigsaws" in *Multiplex and Multilevel Networks*, edited by Battiston, S., Caldarelli, G., and Garas, A. © Oxford University Press 2019.
DOI: 10.1093/oso/9780198809456.003.0002

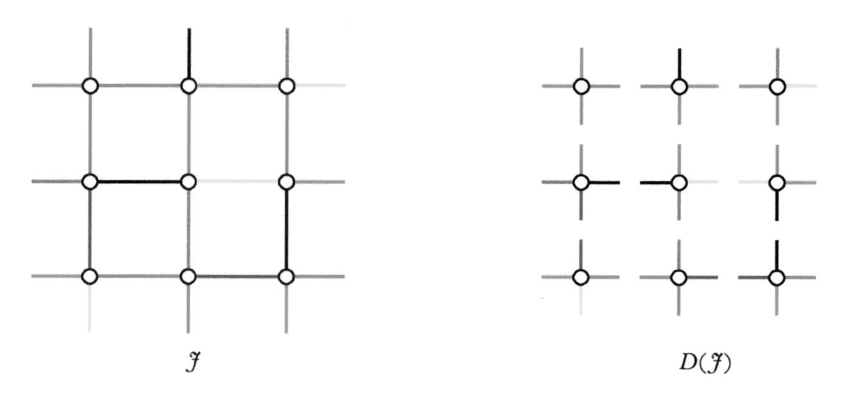


**Figure 2.1** *A (3,8)-jigsaw (left) and its deck (right).*

to the boundary vertices. For $n, q \in \mathbb{N}$, an *$(n,q)$-jigsaw* is a $q$-colored extended $n \times n$ grid, that is, an extended $n \times n$ grid whose $2n(n+1)$ edges are colored using a set of $q$ different colors which we take to be $[q]$ for concreteness (see Figure 2.1). The *tile* of an $(n,q)$-jigsaw corresponding to a vertex $v \in [n]^2$ is given by the coloring of the four edges incident to $v$; more precisely, writing $e_1 = (0,1), e_2 = (1,0), e_3 = -e_1$, and $e_4 = -e_2$, if the edge between $v$ and $v + e_i$ gets color $c_i \in [q]$ for $1 \le i \le 4$, then the tile corresponding to $v$ is the tuple $(c_i)_{i=1}^4 \in [q]^4$. Finally, the *deck* of an $(n,q)$-jigsaw is the multiset of the tiles of the jigsaw, one for each vertex of $[n]^2$.

We now define what it means for a jigsaw to be reconstructible from its deck. Writing $\mathcal{J}(n,q)$ for the set of all $(n,q)$-jigsaws and $\mathcal{D}(n,q)$ for the family of all multisets of size $n^2$ whose elements are chosen from $[q]^4$, let $D\colon \mathcal{J}(n,q) \to \mathcal{D}(n,q)$ be the map sending a jigsaw $\mathcal{J}$ to its deck $D(\mathcal{J})$. We say that a jigsaw $\mathcal{J} \in \mathcal{J}(n,q)$ is *reconstructible* if $D^{-1}(D(\mathcal{J})) = \{\mathcal{J}\}$; equivalently, a jigsaw $\mathcal{J}$ is reconstructible if $D(\mathcal{J}) = D(\mathcal{J}')$ implies $\mathcal{J} = \mathcal{J}'$.

We view $\mathcal{J}(n,q)$ as a probability space by endowing it with a uniform distribution and write $\mathcal{J}(n,q)$ for a random $(n,q)$-jigsaw drawn from this distribution; equivalently, $\mathcal{J}(n,q)$ is a random $(n,q)$-jigsaw generated by independently coloring each edge of the extended $n \times n$ grid with a randomly chosen element of $[q]$. Our primary concern is the following problem about the reconstructibility of a random $(n,q)$-jigsaw, originally raised by Mossel and Ross [197]: of course, there exists only one $(n,1)$-jigsaw for each $n \in \mathbb{N}$ (and this jigsaw is trivially reconstructible), so in what follows, we assume implicitly that $q \ge 2$.

**Problem 2.1.1** *For what $q = q(n)$ is $\mathcal{J}(n,q)$ reconstructible with high probability?*

From below, Mossel and Ross [197] showed that $\mathbb{P}(\mathcal{J}(n,q)$ is reconstructible$) \to 0$ when $q = o(n^{2/3})$, due to the presence of local obstacles to reconstruction: in this regime, a random $(n,q)$-jigsaw contains, with high probability, two configurations, each consisting of two neighboring vertices which may be "exchanged" in the jigsaw, and this is easily seen to obstruct unique reconstruction; however, this argument does not extend to configurations involving single exchangeable vertices (and to a corresponding bound when $q = o(n)$), since the presence of two identical tiles in the deck does not necessarily prevent unique reconstruction. From above, Bordenave, Feige, and Mossel [44] and Nenadov, Pfister, and Steger [207] independently showed, for any fixed $\varepsilon > 0$, that $\mathbb{P}(\mathcal{J}(n,q)$ is reconstructible$) \to 1$ when $q \geq n^{1+\varepsilon}$. Here, we improve on both of these bounds and prove the following, nearly optimal result.

**Theorem 2.1.2** *There exist absolute constants $C, c > 0$ such that as $n \to \infty$, we have*

$$\mathbb{P}(\mathcal{J}(n,q) \textit{ is reconstructible}) \to \begin{cases} 1 & \textit{if } q \geq Cn, \textit{ and} \\ 0 & \textit{if } 2 \leq q \leq cn. \end{cases}$$

The two results contained in the statement of Theorem 2.1.2 are proved by very different methods: the "0-statement" follows from a double-counting argument, while the proof of the "1-statement" is based on an isoperimetric argument which draws from (but is somewhat more involved than) the strategy used by Bordenave, Feige, and Mossel [44] where one attempts to reconstruct a suitably large neighborhood of a tile in order to identify its neighbors in the jigsaw.

We shall prove Theorem 2.1.2 with $C = 10^{40}$, and $c = 1/\sqrt{e}$. With some more effort, it should be possible to refine our proof of Theorem 2.1.2 to show that the result holds for any $C > 1$ (at which point our argument breaks down); however, we choose not to present the details of this stronger claim because we believe the critical number of colors for an $n \times n$ grid to be $n/\sqrt{e}$ and conjecture that the 0-statement in Theorem 2.1.2 is sharp.

**Conjecture 2.1.3** *For any $\varepsilon > 0$, as $n \to \infty$, we have*

$$\mathbb{P}(\mathcal{J}(n,q) \textit{ is reconstructible}) \to 1$$

*for all $q \geq (1/\sqrt{e} + \varepsilon)n$.*

This chapter is organized as follows. We begin with some notation and preliminary discussion in Section 2.2. We give the short proof of the 0-statement in Theorem 2.1.2 in Section 2.3. We prove the key estimate required for the proof of the 1-statement in Theorem 2.1.2 in Section 2.4, and complete the proof of our main result in Section 2.5. We conclude with some discussion in Section 2.6.

**Remark** After we proved the results in this chapter (in November 2016) and were in the process of polishing this write-up, Martinsson [185], working independently, also announced (in January 2017) a proof of a result analogous to Theorem 2.1.2 in a very closely related model (and with a more reasonable constant in the 1-statement). We briefly point out that while the respective 0-statements are established in essentially the same fashion both here and in Ref. [185], the estimates needed to prove the respective 1-statements are established by quite different approaches.

## 2.2 Preliminaries

For a pair of integers $a \le b$, we write $[a,b]$ for the set $\{a, a+1, \ldots, b\}$, and for a natural number $n \in \mathbb{N}$, we abbreviate the set $[1,n]$ by $[n]$.

We define the vectors $e_1 = (0,1), e_2 = (1,0), e_3 = -e_1$, and $e_4 = -e_2$, and we endow the square lattice $\mathbb{Z}^2$ with the graph structure of the infinite grid where two vertices $u, v \in \mathbb{Z}^2$ are adjacent if $u - v = e_i$ for some $1 \le i \le 4$; also, we write $\Lambda$ for the set of edges of the infinite grid on $\mathbb{Z}^2$.

Let $X \subset \mathbb{Z}^2$ be a finite subset of the square lattice. We write $\Lambda(X) \subset \Lambda$ for the set of edges of the grid induced by $X$, and $\partial X \subset \Lambda$ for the *boundary* of $X$, that is, the set of edges between $X$ and $\mathbb{Z}^2 \setminus X$; also, we write $\bar{\Lambda}(X) = \Lambda(X) \cup \partial X$ for the set of edges of the grid with at least one endpoint in $X$. Since $X$ is finite, note that $\mathbb{Z}^2 \setminus X$ contains a unique infinite connected component; the *external boundary* of $X$, written $\partial_e X$, is the set of edges between $X$ and this infinite component, and the *internal boundary* of $X$, written $\partial_i X$, is defined to be $\partial X \setminus \partial_e X$. Finally, the *vertex boundary* of $X$ is defined to be the set of vertices of $X$ incident to some edge of $\partial_e X$.

Observe that if the points of a finite set $X \subset \mathbb{Z}^2$ have $a$ different $x$-coordinates and $b$ different $y$-coordinates in total, then $|X| \le ab$, and the external boundary of $X$ has size at least $2a + 2b$; this observation implies the following well-known isoperimetric statement.

**Proposition 2.2.1** *For any finite set $X \subset \mathbb{Z}^2$, we have $|\partial X| \ge |\partial_e X| \ge 4|X|^{1/2}$.* □

We say that a finite set $X \subset \mathbb{Z}^2$ is *connected* if it is connected when viewed as a subset of the vertex set of the infinite grid, and in what follows, the distance between two points $u, v \in \mathbb{Z}^2$ will always mean the graph distance between $u$ and $v$ in the infinite grid. Also, we say that a finite set of edges $A \subset \Lambda$ is *dual connected* if the corresponding set of edges in the planar dual of the infinite grid is connected. Finally, for $X \subset \mathbb{Z}^2$ and $A \subset \Lambda$, we write $D(X,A)$ for the graph on $X$ whose edge set is $\Lambda(X) \setminus A$; in other words, $D(X,A)$ is the graph induced by $X$ in the grid after we delete the edges in $A$.

It will be convenient to have some notation to deal with maps from $\mathbb{Z}^2$ to $\mathbb{Z}^2$. Let $f$ be an injective map from a finite set $X \subset \mathbb{Z}^2$ to $\mathbb{Z}^2$. We say that a set $Y \subset X$ is *$f$-rigid* if $f(x) - f(y) = x - y$ for all $x, y \in Y$. A *block* of $f$ is a maximal connected rigid subset of $X$; it is easy to check that each vertex of $X$ belongs to a unique block, so the blocks of $f$ partition $X$. An edge $z \in \Lambda(X)$ is said to be *$f$-split* if the endpoints of $z$ belong to different blocks of $f$. We write $\Lambda_f \subset \Lambda$ for the union of $\partial X$ and the set of $f$-split edges; loosely speaking, $\Lambda_f$ is the set of those edges across which we cannot "control" $f$. Note that $\Lambda_f$ may be decomposed into dual-connected components; the following geometric fact about such components will prove useful.

**Proposition 2.2.2** *Let $f$ be an injective map from a finite set $X \subset \mathbb{Z}^2$ to $\mathbb{Z}^2$, and let $A$ be a dual-connected component of $\Lambda_f$. If $Y \subset X$ is a connected component of $D(X,A)$, then the vertices of $Y$ incident to some edge of $A$ are all contained in a single block of $f$.* □

We will also need the following property of finite grids.

**Proposition 2.2.3** *For $n \in \mathbb{N}$, if $X \subset [n]^2$ is such that the distance between any pair of distinct vertices in $X$ is at least three, then $[n]^2 \setminus X$ is connected.* □

Finally, let us quickly restate the problem at hand formally. Note that the edges of the extended $n \times n$ grid are precisely the elements of the set $\bar{\Lambda}([n]^2)$, so an $(n,q)$-jigsaw $\mathcal{J}$ is a map $\mathcal{J}\colon \bar{\Lambda}([n]^2) \to [q]$. Given an $(n,q)$-jigsaw $\mathcal{J}$, the tile $\mathcal{J}_v$ corresponding to a vertex $v \in [n]^2$ is the sequence $(\mathcal{J}(v, v+e_i))_{i=1}^4 \in [q]^4$, and the deck $D(\mathcal{J})$ of $\mathcal{J}$ is the multiset $\{\mathcal{J}_v : v \in [n]^2\}$. As defined previously, a jigsaw $\mathcal{J}$ is reconstructible from its deck if $D(\mathcal{J}') = D(\mathcal{J})$ implies that $\mathcal{J}' = \mathcal{J}$. We write $\mathcal{J}(n,q)$ to denote a random $(n,q)$-jigsaw generated by independently coloring each edge of $\bar{\Lambda}([n]^2)$ with a randomly chosen element of $[q]$. In this language, our primary concern is the following question: for what $q = q(n)$ is $\mathcal{J}(n,q)$ reconstructible with high probability?

We shall make use of standard asymptotic notation; in what follows, the variable tending to infinity will always be $n$ unless we explicitly specify otherwise. We use the term *with high probability* to mean with probability tending to 1 as $n \to \infty$. For the sake of clarity of presentation, we systematically omit floor and ceiling signs whenever they are not crucial.

## 2.3 Proof of the 0-statement

In this short section, we prove the 0-statement in Theorem 2.1.2 by an elementary counting argument.

***Proof of the 0-statement in Theorem 2.1.2*** Recall that $\mathcal{J}(n,q)$ is the set of all $(n,q)$-jigsaws, $\mathcal{D}(n,q)$ is the family of all multisets of size $n^2$ whose elements are chosen from $[q]^4$, and $D\colon \mathcal{J}(n,q) \to \mathcal{D}(n,q)$ is the map sending a jigsaw $\mathcal{J}$ to its deck $D(\mathcal{J})$.

Let $\mathcal{J}_R(n,q) \subset \mathcal{J}(n,q)$ denote the set of all reconstructible jigsaws, that is, jigsaws $\mathcal{J}$ such that $D^{-1}(D(\mathcal{J})) = \{\mathcal{J}\}$. Since $D\colon \mathcal{J}_R(n,q) \to \mathcal{D}(n,q)$ is an injection, $|\mathcal{J}_R(n,q)| \le |\mathcal{D}(n,q)|$. Consequently, we have

$$\mathbb{P}(\mathcal{J}(n,q) \text{ is reconstructible}) = |\mathcal{J}_R(n,q)|/|\mathcal{J}(n,q)| \le |\mathcal{D}(n,q)|/|\mathcal{J}(n,q)|.$$

Now, it is easy to see that

$$|\mathcal{D}(n,q)| = \binom{n^2+q^4-1}{n^2} \quad \text{and} \quad |\mathcal{J}(n,q)| = q^{2n(n+1)},$$

so it follows that

$$\mathbb{P}(\mathcal{J}(n,q) \text{ is reconstructible}) \le \binom{n^2+q^4-1}{n^2-1} q^{-2n^2-2n} \le \binom{n^2+q^4}{n^2} q^{-2n^2-2n}.$$

If $2 \le q \le \sqrt{n}$, then we have

$$\binom{n^2+q^4}{n^2} q^{-2n^2-2n} \le \binom{2n^2}{n^2} 2^{-2n^2-2n} \le 2^{-2n}.$$

If $\sqrt{n} < q \leq n/\sqrt{e}$, on the other hand, then we deduce, using Stirling's approximation, that

$$\binom{n^2+q^4}{n^2} q^{-2n^2-2n} = \frac{q^{2n^2-2n}}{(n^2)!} \prod_{i=1}^{n^2} \left(1 + \frac{i}{q^4}\right) \leq \frac{q^{2n^2-2n}}{(n^2)!} \left(1 + \frac{n^2}{q^4}\right)^{n^2}$$
$$= O\left(\frac{q^{-2n}}{n} \exp\left(n^2 \log\left(\frac{q^2}{n^2}\right) + \frac{n^4}{q^4} + n^2\right)\right) = o\left(q^{-2n}\right).$$

We conclude from the above estimates that

$$\mathbb{P}(\mathcal{J}(n,q) \text{ is reconstructible}) = o(1)$$

for all $2 \leq q \leq n/\sqrt{e}$. □

## 2.4 Towards the 1-statement: Reconstructing large neighborhoods

The starting point of our approach to proving the 1-statement in Theorem 2.1.2 is the strategy adopted by Bordenave, Feige, and Mossel [44] to show that $\mathcal{J} = \mathcal{J}(n,q)$ is reconstructible with high probability when $q \geq n^{1+\varepsilon}$ for some fixed $\varepsilon > 0$. Given the deck $D(\mathcal{J})$ of $\mathcal{J}$, Bordenave, Feige, and Mossel use the following procedure to identify the neighbors of a given tile $\mathcal{J}_v$ with $v \in [n]^2$. For some large integer $k \approx 1/\varepsilon$, they consider all subsets of $D(\mathcal{J})$ of size $(2k+1)^2$ that include the tile $\mathcal{J}_v$, and for each such set, they check if the tiles in that set can be "legally assembled" on a $(2k+1) \times (2k+1)$ grid with $\mathcal{J}_v$ at the center of this grid. While there might exist many such legal assemblies with $\mathcal{J}_v$ at the center, they show that, with high probability, the four neighbors of $\mathcal{J}_v$ in any such legal assembly are identical to the four tiles neighboring $\mathcal{J}_v$ in the original jigsaw. This allows them to identify the neighbors of all tiles corresponding to vertices at distance at least $k$ from the boundary of the grid; once this has been accomplished, it is reasonably straightforward to reconstruct $\mathcal{J}$.

We adopt a similar strategy to the one described above, although in order to show that $\mathcal{J}(n,q)$ is reconstructible when $q \approx n$ (as opposed to when $q \geq n^{1+\varepsilon}$), we require more delicate arguments; for example, we need to take $k \approx \log n$ (as opposed to $k \approx 1/\varepsilon$) and this in turn necessitates more careful estimates.

We now fix positive integers $n, q \in \mathbb{N}$ and set $k = k(n) = \lceil \log n \rceil$; all inequalities in the sequel will hold, provided $n$ and $k$ are sufficiently large.

### 2.4.1 Constraint graphs

Let $\mathcal{J}\colon \bar{\Lambda}([n]^2) \to [q]$ be an $(n,q)$-jigsaw, and let $f$ be an injection from a finite set $X \subset \mathbb{Z}^2$ to $[n]^2$. We say that $f$ *is feasible for* $\mathcal{J}$ if for any pair of adjacent vertices $x, y \in X$, we have $\mathcal{J}(x', x' + y - x) = \mathcal{J}(y', y' + x - y)$, where $x' = f(x)$, and $y' = f(y)$. Clearly, any injective function $f$ as above describes an arrangement of a subset of the tiles of $\mathcal{J}$ on the grid at the vertices of $X$ (where the tile placed at a position $x \in X$ is precisely $\mathcal{J}_{f(x)}$); our definition of feasibility makes precise the notion of when $f$ describes a legal arrangement of tiles. Constraint graphs provide us with an alternate description of legal arrangements, and we define these objects below.

The *constraint graph* of an injective map $f$ from a finite set $X \subset \mathbb{Z}^2$ to $\mathbb{Z}^2$, denoted by $\mathcal{G}_f$, is a graph whose vertex set is a subset of $\Lambda$ and whose edge set contains one edge, called a *constraint*, for each $f$-split edge, where if $\{x,y\}$ is an $f$-split edge with $y = x + e_i$ for some $1 \le i \le 4$, then the constraint corresponding to this edge is an edge joining $\{f(x), f(x) + e_i\}$ and $\{f(y), f(y) - e_i\}$ in the constraint graph; the vertex set of $\mathcal{G}_f$ is the subset of $\Lambda$ spanned by the edges of $\mathcal{G}_f$. In the language of constraint graphs, it is clear that if $\mathcal{J}\colon \bar{\Lambda}([n]^2) \to [q]$ is an $(n,q)$-jigsaw and $f$ is an injection from a finite subset of $\mathbb{Z}^2$ to $[n]^2$, then $f$ is feasible for $\mathcal{J}$ if and only if $\mathcal{J}$ is constant on each connected component of $\mathcal{G}_f$. We define $\gamma(f)$ to be the difference between the size of the vertex set of $\mathcal{G}_f$ and the number of connected components of $\mathcal{G}_f$. We require the following observation due to Bordenave, Feige, and Mossel [44]; we include the short proof for completeness.

**Proposition 2.4.1** *For any injective map $f$ from a finite subset of $\mathbb{Z}^2$ to $[n]^2$, we have*

$$\mathbb{P}(f \textit{ is feasible for } \mathcal{J}(n,q)) = q^{-\gamma(f)}.$$

***Proof*** First, choose a representative from each connected component of $\mathcal{G}_f$. It is clear that $f$ is feasible for $\mathcal{J} = \mathcal{J}(n,q)$ if and only if the following holds: for each vertex of $\mathcal{G}_f$, the color assigned by $\mathcal{J}$ to this vertex is equal to the color assigned by $\mathcal{J}$ to the representative vertex from the corresponding connected component of $\mathcal{G}_f$. Thus, the event that $f$ is feasible for $\mathcal{J}$ is an intersection of $\gamma(f)$ independent events, and each of these events has probability $1/q$; the claim follows. □

It is easy to see that the maximum degree of a constraint graph is, at most, 2, so every constraint graph is a union of paths and cycles; this observation implies the following.

**Proposition 2.4.2** *If $f$ is an injection from a finite subset of $\mathbb{Z}^2$ to $\mathbb{Z}^2$, then $\gamma(f) \ge |V(\mathcal{G}_f)|/2 \ge |E(\mathcal{G}_f)|/2$.* □

## 2.4.2 Windows

To make precise the idea of recovering the four tiles neighboring a given tile by attempting to reconstruct the large neighborhood of the tile in question, we need the notion of a "window."

For $v \in [n]^2$ and an $(n,q)$-jigsaw $\mathcal{J}$, a *$v$-window* with respect to $\mathcal{J}$ is an injective map $f\colon [-k,k]^2 \to [n]^2$ such that $f(0,0) = v$, and $f$ is feasible for $\mathcal{J}$; we remind the reader that $k = \lceil \log n \rceil$ here, and in what follows.

If $v \in [n]^2$ is at distance at least $k$ from the vertex boundary of the $n \times n$ grid, then the map defined by $f(x) = v + x$ for all $x \in [-k,k]^2$ is a $v$-window; more generally, if there exists some $v' \in [n]^2$ at distance at least $k$ from the vertex boundary of the $n \times n$ grid such that $\mathcal{J}_{v'} = \mathcal{J}_v$, then the map defined by $f(0,0) = v$ and $f(x) = v' + x$ for all $x \in [-k,k]^2 \setminus \{(0,0)\}$ is a $v$-window. A $v$-window $f$ is said to be *trivial* if $(\mathcal{J}_{f(e_i)})_{i=1}^4 = (\mathcal{J}_{v'+e_i})_{i=1}^4$ for some $v' \in [n]^2$ such that $\mathcal{J}_{v'} = \mathcal{J}_v$; in other words, a $v$-window is trivial if the four tiles neighboring $\mathcal{J}_v$ in the $v$-window are identical to the four tiles neighboring some tile $\mathcal{J}_{v'}$ in the jigsaw, with $\mathcal{J}_{v'}$ itself identical to $\mathcal{J}_v$. This definition of triviality is motivated by the fact that when $q \approx n$, the deck of $\mathcal{J}(n,q)$ may contain some tiles of multiplicity greater than 1 (although, as we shall see, this will not present an obstacle to reconstruction). We shall show, provided $q$ is suitably large, that all windows with respect to $\mathcal{J}(n,q)$ are trivial with high probability; the aim of this section is to establish the following lemma.

**Lemma 2.4.3** *If $q \geq 10^{40}n$, then $\mathcal{J}(n,q)$ has the following property with high probability: for each $v \in [n]^2$, every $v$-window with respect to $\mathcal{J}(n,q)$ is trivial.*

### 2.4.3 Templates

To prove Lemma 2.4.3, it is natural to first attempt to use a union bound over all candidate injective maps from $[-k,k]^2$ to $[n]^2$; however, this turns out to be too crude for our purposes. The reason for this is roughly as follows: the number of candidate windows is artificially inflated by maps $f\colon [-k,k]^2 \to [n]^2$ with a large number of "holes"; more precisely, there exist too many candidate windows $f\colon [-k,k]^2 \to [n]^2$ with the property that one of the blocks of $f$ is contained entirely in the interior of another block of $f$. One could hope to address this issue by locally modifying a candidate window so as to remove such pairs of "nested blocks," but attempting to do so results in a situation where some tiles of the jigsaw end up getting used multiple times.

To circumvent the difficulties outlined above, we introduce the notion of a "template." To introduce this notion, it will be helpful to first have some notation.

Let $A \subset \bar{\Lambda}([-k,k]^2)$ be a set of edges of the grid. Recall that $D([-k,k]^2, A)$ is the graph on $[-k,k]^2$ whose edge set is $\Lambda([-k,k]^2) \setminus A$. For any connected component $X \subset [-k,k]^2$ of $D([-k,k]^2, A)$, we define the *quasiblock* $\hat{X}$ associated with $X$ to be the set of vertices of $X$ incident to some edge in $A$; in the sequel, when we refer to a quasiblock $\hat{X}$ of $A$, we implicitly assume that the corresponding connected component of $D([-k,k]^2, A)$ is denoted by $X$. Finally, we write $A^*$ for the set $A \cap \Lambda([-k,k]^2)$.

For $v \in [n]^2$, a *$v$-template* is a pair $(A,h)$, where $A \subset \bar{\Lambda}([-k,k]^2)$ and $h$ is an injective map from the union of the quasiblocks of $A$ to $[n]^2$, such that

1. $A$ contains at least one edge incident to $(0,0)$,
2. $A$ does not consist of precisely the four edges incident either to $(0,0)$ or to one of its four neighbors,
3. $A$ is dual connected,
4. $h(0,0) = v$,
5. either $\partial[-k,k]^2 \subset A$ or $\partial[-k,k]^2 \cap A = \varnothing$,
6. each quasiblock of $A$ is $h$-rigid, and
7. each edge of $A^*$ is $h$-split.

Given an $(n,q)$-jigsaw $\mathcal{J}$, we abuse notation slightly and say that a $v$-template $(A,h)$ is feasible for $\mathcal{J}$ if $h$ is feasible for $\mathcal{J}$. The definition of a template is motivated by the following fact.

**Proposition 2.4.4** *Let $\mathcal{J}$ be an $(n,q)$-jigsaw, and let $v \in [n]^2$. If there exists a nontrivial $v$-window $f$ with respect to $\mathcal{J}$, then there exists a $v$-template $(A,h)$ that is feasible for $\mathcal{J}$.*

*Proof* Since any tile is uniquely determined by its four neighbors in any valid arrangement of tiles, it is easy to check using the fact that $f$ is a nontrivial $v$-window that there exists an $f$-split edge $z$ incident to $(0,0)$ with the property that the dual-connected component of $z$ in $\Lambda_f$ does not consist of precisely the four edges incident either to $(0,0)$ or to one of its four neighbors. We now take $A$ to be the dual-connected component of $z$ in $\Lambda_f$, and $h$ to be the restriction of $f$ to the endpoints of $A$ in $[-k,k]^2$.

Clearly, $A$ contains at least one edge incident to $(0,0)$, does not consist of precisely the four edges incident either to $(0,0)$ or to one of its four neighbors, and is dual connected. As $f$ is a $v$-window that extends $h$, we have $\mathrm{h}(0,0) = \mathrm{v}$. Next, since $A$ is a dual-connected component of $\Lambda_f$ and $\partial[-k,k]^2$ is a dual-connected subset of $\Lambda_f$, either $\partial[-k,k]^2 \subset A$, or $\partial[-k,k]^2 \cap A = \varnothing$. Furthermore, it follows from Proposition 2.2.2 that every quasiblock of $A$ is a subset of a single block of $f$; since $f$ extends $h$, it follows that every quasiblock of $A$ is $h$-rigid. Finally, since each edge of $A^*$ is $f$-split, each edge of $A^*$ must also be $h$-split. □

We shall prove Lemma 2.4.3, using a union bound over templates as opposed to windows; in particular, we shall show, provided $q$ is suitably large, that, with sufficiently high probability, no $v$-template $(A,h)$ is feasible for $\mathcal{J}(n,q)$.

We say that a template $(A,h)$ is *large* if $\partial[-k,k]^2 \subset A$, and *small* if $A \cap \partial[-k,k]^2 = \varnothing$. Of course, every template is either large or small. We shall require slightly different arguments to deal with large and small templates. The following fact will prove useful when estimating the number of templates of both types; see Problem 45 in [41], for instance.

**Proposition 2.4.5** *In a graph of maximal degree $\Delta$, the number of connected induced subgraphs with $l+1$ vertices, one of which is a given vertex, is at most $(e(\Delta-1))^l$.* □

### 2.4.4 Large templates

We will need an estimate for the number of large templates, as well as an estimate for the probability that such a template is feasible for $\mathcal{J}(n,q)$. In order to simplify our bookkeeping, it will be helpful to introduce the notion of a "cluster." Let $(A,h)$ be a large $v$-template. For a quasiblock $\hat{X} \subset [-k,k]^2$ of $A$, let $h(\hat{X}) \subset [n]^2$ denote the (rigid) image of $\hat{X}$ under $h$. Let us define the *cluster graph* of $(A,h)$ to be the graph on the quasiblocks of $A$ where two quasiblocks $\hat{X}$ and $\hat{Y}$ are adjacent if there exists an edge of the lattice between $h(x)$ and $h(y)$ for some $x \in \hat{X}$ and $y \in \hat{Y}$ and, furthermore, this edge belongs to the external boundary of both $h(\hat{X})$ and $h(\hat{Y})$. A *cluster* of $(A,h)$ is then a subset of $[n]^2$ consisting of the images of all the quasiblocks in a connected component of the cluster graph.

For nonnegative integers $\delta$, $r_1$, and $r_2$, we say that a large $v$-template $(A,h)$ is of type $(\delta,r_1,r_2)$ if $|A|=\delta$, the number of quasiblocks of $A$ is $r_1+r_2$, and the number of clusters of $(A,h)$ is $r_1$. Writing $N_l(\delta,r_1,r_2)$ for the number of large $v$-templates of type $(\delta,r_1,r_2)$, we have the following estimate.

**Proposition 2.4.6** *For nonnegative integers $\delta$, $r_1$, and $r_2$, we have*

$$N_l(\delta,r_1,r_2)=\begin{cases}0 & \text{if } \delta<8k+4 \text{ or } \delta<r_1+r_2, \text{ and}\\ O(30^{\delta}n^{2r_1}k^{6r_2}/n^2) & \text{otherwise.}\end{cases}$$

***Proof*** We estimate the number of large $v$-templates $(A,h)$ of type $(\delta,r_1,r_2)$ by first estimating the number of ways in which we may choose $A$ and then estimating the number of ways in which we may choose $h$ once we are given $A$.

First, we may assume that $\delta\geq 8k+4$ since if $(A,h)$ is a large $v$-template, then $\partial[-k,k]^2\subset A$ by definition. Second, we may also suppose that $\delta\geq r_1+r_2$; indeed, by considering a northernmost vertex of each quasiblock of $A$, for example, we observe that the number of quasiblocks of $A$ is, at most, the size of $A$, so the claimed bound holds trivially in the case where $\delta<r_1+r_2$.

We now estimate the number of ways to choose $A$. Since $A$ must contain an edge incident to (0,0) and must additionally be dual connected, it follows from Proposition 2.4.5 that the number of choices for $A$ (even ignoring the restriction that $A$ has precisely $r_1+r_2$ quasiblocks) is at most $4(5e)^{\delta-1}\leq 15^{\delta}$, as each edge of the square lattice is adjacent to six other edges of the square lattice in the planar dual of the lattice.

Next, we estimate the number of ways to choose $h$ for a given $A$. Once we fix an $A$ with $r_1+r_2$ quasiblocks, it suffices to specify the image of one vertex from each quasiblock of $A$ under $h$ to completely specify $h$, since each quasiblock of $A$ is $h$-rigid. We count the number of ways to choose $h$ as follows. We first choose $r_1$ *representative* quasiblocks in such a way that these quasiblocks all belong to different clusters, while ensuring that the quasiblock containing (0,0) is one of these representatives; the number of ways to choose these representatives is at most

$$\binom{r_1+r_2-1}{r_1-1}\leq 2^{r_1+r_2}\leq 2^{\delta}.$$

Of course, since $h(0,0)=v$, this specifies the image of the quasiblock containing (0,0). We then specify the image of a vertex (say, the northernmost) from each of the remaining $r_1-1$ representative quasiblocks; this may be done in $n^{2(r_1-1)}$ ways. Finally, we note that there are $O(k^6)$ choices for the image of one of the $r_2$ leftover quasiblocks. To see this, note that each leftover quasiblock belongs to the same cluster as one of the representative quasiblocks, so the image of such a leftover quasiblock must be at distance at most $(2k+1)^2$ from the image of one of the representative quasiblocks; the claimed bound follows, since there are at most $(2k+1)^2$ points contained in the representative quasiblocks and there are at most $(2d+1)^2$ points at distance at most $d$ from any fixed point of the grid. Combining these estimates, we see that the number of choices for $h$ once we have specified $A$ is $O(2^\delta n^{2(r_1-1)} k^{6r_2})$.

It now follows that

$$N(\delta, r_1, r_2) = O(15^\delta 2^\delta n^{2(r_1-1)} k^{6r_2}) = O(30^\delta n^{2r_1} k^{6r_2}/n^2). \qquad \square$$

To estimate the probability that a large $v$-template $(A,h)$ is feasible for $\mathcal{J}(n,q)$, we shall appeal to Proposition 2.4.1, which gives us a bound for this probability in terms of $\gamma(h)$; recall that $\gamma(h)$ is the difference between the size of the vertex set of $\mathcal{G}_h$ and the number of connected components of $\mathcal{G}_h$, where $\mathcal{G}_h$ is the constraint graph of $h$.

**Proposition 2.4.7** *If $(A,h)$ is a large $v$-template of type $(\delta, r_1, r_2)$, then we have $\gamma(h) \geq \delta/20$, and $\gamma(h) \geq 2r_1 + r_2/2 - 2r_1/(2k+1)$.*

***Proof*** We shall use Proposition 2.4.2 to bound $\gamma(h)$ from below. We will estimate the size of both the vertex set and the edge set of $\mathcal{G}_h$.

Since $\mathcal{G}_h$ contains one edge for each $h$-split edge, it is easy to see that the edge set of $\mathcal{G}_h$ has size at least $A^*$, so $|E(\mathcal{G}_h)| \geq |A^*| = |A| - (8k+4)$, as $|\partial[-k,k]^2| = 8k+4$. Now, since $A$ contains an edge incident to $(0,0)$, is dual connected, and also contains $\partial[-k,k]^2$, we have $|A^*| \geq k$ and, consequently, $|A| \geq 9k+4$; it follows, provided $k$ is sufficiently large, that $|E(\mathcal{G}_h)| \geq |A| - (8k+4) \geq |A|/10 = \delta/10$. We now conclude from Proposition 2.4.2 that $\gamma(h) \geq |E(\mathcal{G}_h)|/2 \geq \delta/20$.

To estimate the size of the vertex set of $\mathcal{G}_h$, we begin with the following observation. First, if $\hat{X}$ is a quasiblock of $A$, then since $\hat{X}$ is $h$-rigid, there is a one-to-one correspondence between $\partial_e \hat{X}$ and $\partial_e h(\hat{X})$. Next, note that each edge of $\partial_e \hat{X}$ is either an element of $A^*$ (and consequently $h$-split) or an element of $\partial[-k,k]^2$. It now follows that each edge of $\partial_e h(\hat{X})$ that corresponds to an edge of $\partial_e \hat{X}$ contained in $A^*$ must belong to the vertex set of $\mathcal{G}_h$.

For a cluster $K$ of the template $(A,h)$ composed of the images of the quasiblocks $\hat{X}_1, \hat{X}_2, \ldots, \hat{X}_m$, we write $S(K)$ for the set of edges between $\hat{X}_i$ and $\hat{X}_j$ for some $1 \leq i < j \leq m$, and $T(K)$ for the set $S(K) \cup \partial_e K$. First, it is clear that $S(K)$ and $\partial_e K$ are disjoint for each cluster $K$. Furthermore, it is also easy to see that if $K_1$ and $K_2$ are distinct clusters, then the sets $T(K_1)$ and $T(K_2)$ are disjoint. Let $T \subset \bar{\Lambda}([n]^2)$ denote the union of the sets $T(K)$, where $K$ runs over the $r_1$ clusters of $(A,h)$. From our earlier discussion, it follows that an edge of $T$ is a vertex of $\mathcal{G}_h$ unless it corresponds to an edge in $\partial[-k,k]^2$. Consequently, we have $|V(\mathcal{G}_h)| \geq |T| - (8k+4)$.

We now use an isoperimetric argument to bound $|T|$ from below; we begin with following observation.

**Claim 2.4.8** *For a cluster $K$ of $(A,h)$ composed of the images of the quasiblocks $\hat{X}_1$, $\hat{X}_2, \ldots, \hat{X}_m$, we have*

$$|T(K)| \geq 4\sqrt{|X_1|+|X_2|+\cdots+|X_m|}+m-1.$$

***Proof*** It immediately follows from the fact that $K$ corresponds to a connected component of size $m$ in the cluster graph of $(A,h)$ that $|S(K)| \geq m-1$. Next, while Proposition 2.2.1 immediately tells us that

$$|\partial_e K| \geq 4\sqrt{|\hat{X}_1|+|\hat{X}_2|+\cdots+|\hat{X}_m|},$$

we may get a better estimate as follows. Note that since $\partial[-k,k]^2 \subset A$, the quasiblock $\hat{X}$ of $A$ associated with a connected component $X$ of $D([-k,k]^2,A)$ is, in fact, the vertex boundary of $X$. Therefore, it follows from the Jordan curve theorem that each quasiblock of $A$ must divide the plane into an exterior and an interior region. From the definition of a cluster, it follows that $h(\hat{X}_i)$ lies in the exterior of $h(\hat{X}_j)$ for all $i \neq j$. Consequently, it follows that $\partial_e K$ is, in fact, the external boundary of a set of size $|X_1|+|X_2|+\cdots+|X_m|$; therefore, we have

$$|\partial_e K| \geq 4\sqrt{|X_1|+|X_2|+\cdots+|X_m|}.$$

The claim follows since $S(K)$ and $\partial_e K$ are disjoint. □

By summing the bound from Proposition 2.4.8 over the $r_1$ clusters of $(A,h)$, we obtain a bound of the form

$$|T| \geq 4\sqrt{a_1}+4\sqrt{a_2}+\cdots+4\sqrt{a_{r_1}}+r_2$$

for some collection of positive integers $a_1,a_2,\ldots,a_{r_1}$ satisfying $a_1+a_2+\cdots+a_{r_1} = (2k+1)^2$; this is immediate once we note that each connected component of $D([-k,k]^2,A)$ contributes precisely once to the bound in Proposition 2.4.8 as we run over the clusters of $(A,h)$. We conclude, using convexity, that

$$|T| \geq 4(r_1-1)+4\sqrt{(2k+1)^2-(r_1-1)}+r_2 \geq 4r_1+r_2+8k+4-\frac{4r_1}{(2k+1)}.$$

We know from Proposition 2.4.2 that $\gamma(h) \geq |V(\mathcal{G}_h)|/2 \geq |T|/2-(4k+2)$; it now follows that $\gamma(h) \geq 2r_1+r_2-2r_1/(2k+1)$. □

## 2.4.5 Small templates

We shall handle small templates using arguments similar to those used to deal with large templates; however, some small subtleties necessitate a slightly different approach to bookkeeping. If $(A,h)$ is small $v$-template, then it may well be the case that $|A|$ is small, so our estimates need to be capable of handling this; this cannot happen when $(A,h)$ is large, since $\partial[-k,k]^2 \subset A$ in this case. On the other hand, if $(A,h)$ is small, then since $\partial[-k,k]^2 \cap A = \varnothing$, we do not need to worry about overcounting contributions from $\partial[-k,k]^2$ when estimating $\gamma(h)$. We will modify the arguments we used to deal with large templates slightly in order to balance these considerations.

Let $(A,h)$ be a small $v$-template. Since $A \cap \partial[-k,k]^2 = \varnothing$, it is easy to verify that the vertex boundary of $[-k,k]^2$ is contained in a single connected component of $D([-k,k]^2,A)$; we call the quasiblock corresponding to this connected component the *boundary quasiblock* of $A$ and refer to the other quasiblocks of $A$ as *non-boundary quasiblocks*.

We will need a slight modification of the notion of a "cluster" that distinguishes between the boundary quasiblock and non-boundary quasiblocks. Let $(A,h)$ be a small $v$-template and as before, for a quasiblock $\hat{X} \subset [-k,k]^2$ of $A$, let $h(\hat{X}) \subset [n]^2$ denote the (rigid) image of $\hat{X}$ under $h$. Let us define the *cluster graph* of $(A,h)$ to be the graph on the quasiblocks of $A$ where

1. two non-boundary quasiblocks $\hat{X}$ and $\hat{Y}$ are adjacent if there exists an edge of the square lattice between $h(x)$ and $h(y)$ for some $x \in \hat{X}$ and $y \in \hat{Y}$ and, furthermore, this edge belongs to the external boundary of both $h(\hat{X})$ and $h(\hat{Y})$, and
2. the boundary quasiblock $\hat{X}$ and a non-boundary quasiblock $\hat{Y}$ are adjacent if there exists an edge of the square lattice between $h(x)$ and $h(y)$ for some $x \in \hat{X}$ and $y \in \hat{Y}$ and, furthermore, this edge belongs to the internal boundary of $h(\hat{X})$ and the external boundary of $h(\hat{Y})$.

A *cluster* of $(A,h)$ is then a subset $[n]^2$ consisting of the images of all the quasiblocks in a connected component of the cluster graph; again, we call the cluster containing the image of the boundary quasiblock the *boundary cluster* and refer to the other clusters as *non-boundary clusters*.

As before, for nonnegative integers $\delta$, $r_1$, and $r_2$, we say that a small $v$-template $(A,h)$ is of type $(\delta, r_1, r_2)$ if $|A| = \delta$, the number of quasiblocks of $A$ is $r_1 + r_2$, and the number of clusters of $(A,h)$ is $r_1$. Writing $N_s(\delta, r_1, r_2)$ for the number of small $v$-templates of type $(\delta, r_1, r_2)$, we have the following estimate, the proof of which is identical to that of Proposition 2.4.6.

**Proposition 2.4.9** *For nonnegative integers $\delta$, $r_1$, and $r_2$, we have*

$$N_s(\delta, r_1, r_2) = O(30^{\delta} n^{2r_1} k^{6r_2} / n^2). \qquad \square$$

To estimate the probability that a small $v$-template $(A,h)$ is feasible for $\mathcal{J}(n,q)$, we will use the following.

**Proposition 2.4.10** *If $(A,h)$ is a small $v$-template of type $(\delta, r_1, r_2)$, then we have $\gamma(h) \geq \delta/2$, and $\gamma(h) \geq 2r_1 + r_2/2 + 1/2$.*

***Proof*** As before, we will estimate the size of both the vertex set and the edge set of the constraint graph $\mathcal{G}_h$.

Since $\mathcal{G}_h$ contains one edge for each $h$-split edge, it is easy to see that edge set of $\mathcal{G}_h$ has size at least $A^*$. Since $\partial[-k,k]^2 \cap A = \varnothing$, we have $A^* = A$, so $|E(\mathcal{G}_h)| \geq |A| = \delta$. We now conclude from Proposition 2.4.2 that $\gamma(h) \geq |E(\mathcal{G}_h)|/2 \geq \delta/2$.

To estimate the size of the vertex set of $\mathcal{G}_h$, we begin with the following observations. First, if $\hat{X}$ is a non-boundary quasiblock of $A$, then since $\hat{X}$ is $h$-rigid, there is a one-to-one correspondence between $\partial_e \hat{X}$ and $\partial_e h(\hat{X})$; since each edge of $\partial_e \hat{X}$ is an element of $A$ (and consequently $h$-split, as $A = A^*$), it follows that each edge of $\partial_e h(\hat{X})$ must belong to the vertex set of $\mathcal{G}_h$. Next, if $\hat{X}$ is the boundary quasiblock of $A$, then since each edge of $\partial_i \hat{X}$ is an element of $A$, it follows that each edge of $\partial_i h(\hat{X})$ must belong to the vertex set of $\mathcal{G}_h$. For a non-boundary cluster $K$ of $(A,h)$ composed of the images of the non-boundary quasiblocks $\hat{X}_1, \hat{X}_2, \dots, \hat{X}_m$, we write $S(K)$ for the set of edges between $\hat{X}_i$ and $\hat{X}_j$ for some $1 \le i < j \le m$, and $T(K)$ for the set $S(K) \cup \partial_e K$; it is clear that $S(K)$ and $\partial_e K$ are disjoint, so $T(K)$ is, in fact, the disjoint union of these sets. For the boundary cluster $K$ composed of the images of the boundary quasiblock $X$ and non-boundary quasiblocks $\hat{X}_1, \hat{X}_2, \dots, \hat{X}_m$, we write $S(K)$ for the set of edges between $\hat{X}_i$ and $\hat{X}_j$ for some $1 \le i < j \le m$, and $T(K)$ for the set $S(K) \cup \partial_i h(\hat{X})$; again, it is clear that $S(K)$ and $\partial_i h(\hat{X})$ are disjoint and that $T(K)$ is the disjoint union of these sets. Finally, it is also easy to see that if $K_1$ and $K_2$ are distinct clusters, then the sets $T(K_1)$ and $T(K_2)$ are disjoint. As before, let $T \subset \bar{\Lambda}([n]^2)$ denote the union of the sets $T(K)$, where $K$ runs over the $r_1$ clusters of $(A,h)$. From our earlier observations, it follows that each edge of $T$ is a vertex of $\mathcal{G}_h$; consequently, we have $|V(\mathcal{G}_h)| \ge |T|$.

To bound $|T|$ from below, we first deal with non-boundary clusters.

**Claim 2.4.11** *For a non-boundary cluster $K$ of $(A,h)$ composed of the images of the non-boundary quasiblocks $\hat{X}_1, \hat{X}_2, \dots, \hat{X}_m$, we have $|T(K)| \ge 4 + (m-1)$.* □

***Proof*** It immediately follows from the fact that $K$ corresponds to a connected component of size $m$ in the cluster graph of $(A,h)$ that $|S(K)| \ge m-1$. Since the external boundary of a non-empty subset of the square lattice contains at least four edges, it follows that $|\partial_e K| \ge 4$. The claim follows since $S(K)$ and $\partial_e K$ are disjoint. □

Next, we have the following estimate for the boundary cluster.

**Claim 2.4.12** *If the boundary cluster $K$ of $(A,h)$ is composed of the images of the boundary quasiblock $\hat{X}$ and non-boundary quasiblocks $\hat{X}_1, \hat{X}_2, \dots, \hat{X}_m$, then $|T(K)| \ge 6 + (m-1)$.*

***Proof*** As before, it is clear that $|S(K)| \ge m-1$. We claim that $|\partial_i h(\hat{X})| \ge 6$. To see this, consider the connected component $X$ of $D([-k,k]^2, A)$ containing the vertex boundary of $[-k,k]^2$. Writing $X' = [-k,k]^2 \setminus X$, note we must have $|X'| \ge 2$, for if not, then $A$ must consist of precisely the four edges incident either to $(0,0)$ or to one of its four neighbors. Note also that $\partial_e X' \subset \partial_i X$. Now, since $X'$ contains at least two vertices, it is easily verified that $\partial_e X'$ contains at least six edges;

consequently, $|\partial_i h(\hat{X})| = |\partial_i \hat{X}| = |\partial_i X| \geq 6$. The claim follows, since $S(K)$ and $\partial_i h(\hat{X})$ are disjoint. □

By summing the bound from Claim 2.4.11 over the $r_1 - 1$ non-boundary clusters of $(A, h)$ and then adding the bound from Claim 2.4.12, we obtain

$$|T| \geq 4(r_1 - 1) + 6 + (r_2 - 1) = 4r_1 + r_2 + 1.$$

We know from Proposition 2.4.2 that $\gamma(h) \geq |V(\mathcal{G}_h)|/2 \geq |T|/2$, so it follows from the above bound that $\gamma(h) \geq 2r_1 + r_2/2 + 1/2$. □

### 2.4.6 Proof of the main lemma

We are now in a position to prove Lemma 2.4.3.

***Proof of Lemma 2.4.3*** We shall show for any $v \in [n]^2$, using a union bound over all $v$-templates, that the probability that there exists a $v$-template that is feasible for $\mathcal{J} = \mathcal{J}(n, q)$ is $o(n^{-2})$ when $q \geq 10^{40} n$; the lemma then follows from a union bound over the elements of $[n]^2$.

Fix a vertex $v \in [n]^2$. Let $E_l$ denote the event that there exists a large $v$-template that is feasible for $\mathcal{J}$, and let $E_s$ denote the event that there exists a small $v$-template that is feasible for $\mathcal{J}$.

First, we bound $\mathbb{P}(E_l)$ as follows. Consider the event $E_l(\delta, r_1, r_2)$ that there exists a large $v$-template of type $(\delta, r_1, r_2)$ that is feasible for $\mathcal{J}$. Of course, Proposition 2.4.6 implies that $\mathbb{P}(E_l(\delta, r_1, r_2)) = 0$ either if $\delta < 8k + 4$ or if $\delta < r_1 + r_2$. Otherwise, from Propositions 2.4.6 and 2.4.7 and the fact that $k = \lceil \log n \rceil$, we see that

$$\mathbb{P}(E_l(\delta, r_1, r_2)) = O\left(\frac{30^\delta n^{2r_1} k^{6r_2}}{n^2 100^\delta n^{2r_1 + r_2/2 - 2r_1/(2k+1)}}\right) = O\left(\frac{30^\delta n^{2r_1/(2k+1)}}{n^2 100^\delta}\right)$$
$$= O\left(\frac{e^{r_1}}{n^2 3^\delta}\right) = O\left(\frac{e^\delta}{n^2 3^\delta}\right) = O\left(\frac{(e/3)^{\log n}}{n^2}\right) = O\left(n^{-1-\log 3}\right).$$

Now, since $1 + \log 3 > 2$ and $k = \lceil \log n \rceil$, we deduce from the above estimate that

$$\mathbb{P}(E_l) = \sum_{\delta=1}^{4(2k+1)^2} \sum_{r_1=1}^{(2k+1)^2} \sum_{r_2=1}^{(2k+1)^2} \mathbb{P}(E_l(\delta, r_1, r_2)) = O\left(4(2k+1)^6 n^{-1-\log 3}\right) = o(n^{-2}).$$

Next, we bound $\mathbb{P}(E_s)$ as follows. Consider the event $E_s(\delta, r_1, r_2)$ that there exists a small $v$-template of type $(\delta, r_1, r_2)$ that is feasible for $\mathcal{J}$. From Propositions 2.4.9 and 2.4.10 and the fact that $k = \lceil \log n \rceil$, we see that

$$\mathbb{P}(E_s(\delta, r_1, r_2)) = O\left(\frac{30^\delta n^{2r_1} k^{6r_2}}{n^2 1020^\delta n^{2r_1 + r_2/2 + 1/2}}\right) = O\left(\frac{30^\delta}{n^{2+1/2} 100^\delta}\right) = O\left(n^{-2-1/2}\right).$$

As before, since $k = \lceil \log n \rceil$, we deduce from the above estimate that

$$\mathbb{P}(E_s) = \sum_{\delta=1}^{4(2k+1)^2} \sum_{r_1=1}^{(2k+1)^2} \sum_{r_2=1}^{(2k+1)^2} \mathbb{P}(E_s(\delta, r_1, r_2)) = O\Big(4(2k+1)^6 n^{-2-1/2}\Big) = o(n^{-2}).$$

It follows that the probability that there exists a $v$-template that is feasible for $\mathcal{J}$ is $o(n^{-2})$; the lemma follows from a union bound over the vertices of the grid. □

## 2.5 Proof of the 1-statement

In this section, we prove the 1-statement in Theorem 2.1.2. We proceed roughly as in Ref. [44] by first assembling the "central bulk" of a random jigsaw using Lemma 2.4.3, and then extending this assembly to the "periphery" in a fairly straightforward fashion; our arguments will, however, require a bit more work than the one in Ref. [44], since we have fewer colors to work with.

***Proof of the 1-statement in Theorem 2.1.2*** Suppose that $q \geq 10^{40} n$, let $\mathcal{J} = \mathcal{J}(n,q)$, and, as in Section 2.4, let $k = \lceil \log n \rceil$. To prove the 1-statement, we shall describe an algorithm that reconstructs $\mathcal{J}$ from its deck $D(\mathcal{J})$ with high probability.

We begin by addressing the possibility of tiles occurring with multiplicity greater than one in $D(\mathcal{J})$. Let $X_1$ denote the number of pairs $(u,v) \in ([n]^2)^2$ with $\mathcal{J}_u = \mathcal{J}_v$, and let $X_2$ denote the number of pairs $(u,v) \in ([n]^2)^2$ with $\mathcal{J}_u = \mathcal{J}_v$ such that $u$ and $v$ are additionally at distance at most two from each other. We then observe the following.

**Claim 2.5.1** $\mathbb{E}[X_1] \leq 1$, *and* $\mathbb{E}[X_2] = o(1)$.

***Proof*** The claim follows immediately from noting that $\mathbb{E}[X_1] = n^4 q^{-4}$ and that $\mathbb{E}[X_2] = O(n^2 q^{-4})$. □

Let us now record some properties that are possessed by $\mathcal{J}$ with high probability:

(A) There exist no nontrivial $v$-windows with respect to $\mathcal{J}$ for any $v \in [n]^2$; this follows from Lemma 2.4.3.

(B) The number of vertices $v \in [n]^2$ such that the tile $\mathcal{J}_v$ has multiplicity greater than one in $D(\mathcal{J})$ is at most $\log n$; this follows from Claim 2.5.1 and Markov's inequality.

(C) If $\mathcal{J}_u = \mathcal{J}_v$ for some $u, v \in [n]^2$, then the distance between $u$ and $v$ is at least 3; this again follows from Claim 2.5.1 and Markov's inequality.

We first show how one may reconstruct a large subgrid of $\mathcal{J}$ from $D(\mathcal{J})$ with high probability; we do this by showing how one may perform this reconstruction assuming that $\mathcal{J}$ satisfies (A), (B), and (C). To this end, we proceed by building a labeled, directed graph $H$ on $D(\mathcal{J})$ to encode the relative positions of the tiles in the jigsaw. In what follows, a component of the directed graph $H$ will mean a connected component of the underlying undirected graph.

First, we consider every tile $t \in D(\mathcal{J})$ that occurs in the deck with multiplicity 1. For such a tile $t$, we consider all possible subsets of $(2k+1)^2$ tiles that include $t$, and for each such set, we consider all possible arrangements of this set of tiles on the grid $[-k,k]^2$ with $t$ being placed at $(0,0)$. Finally, for each such arrangement that is feasible, we record the tuple $(t,t_1,t_2,t_3,t_4)$, where $t_i$ is the tile placed at $e_i$ in this arrangement for $1 \le i \le 4$. Now, for each recorded tuple $(t,t_1,t_2,t_3,t_4)$, we add an edge directed from $t$ to $t_i$ labeled $e_i$ in $H$ if the tile $t_i$ also occurs with multiplicity one in the deck.

It follows from (A) that if there exists a directed edge from a tile $t$ to a tile $t'$ labeled $e_i$ in $H$, then it must be the case that $t = \mathcal{J}_v$ and $t' = \mathcal{J}_{v'}$, where $v, v' \in [n]^2$ are vertices such that $v' = v + e_i$. Consequently, each component of $H$ describes the relative positions of the tiles in that component in $\mathcal{J}$; in other words, for any two tiles $t = \mathcal{J}_v$ and $t' = \mathcal{J}_{v'}$ that belong to the same component in $H$, we may determine $v - v'$ using $H$.

From (C) and Proposition 2.2.3, we deduce that the tiles of $\mathcal{J}$ coming from the central $(n-2k) \times (n-2k)$ subgrid of $[n]^2$ which furthermore appear with multiplicity 1 in $D(\mathcal{J})$ all belong to the same component of $H$; it follows from (B) that this component contains at least $(n-2k)^2 - \log n > n^2/2$ tiles and is consequently the unique largest component of $H$.

Next, we fill in the "holes" in the largest component of $H$ as follows. We know that we may determine, up to translation, the positions on the square lattice of all the tiles in a given component of $H$; we fix an arrangement of the tiles in the largest component by placing one of these tiles at the origin, and the other tiles at their appropriate positions relative to the origin. Suppose that there is no tile at some position $x \in \mathbb{Z}^2$ in this arrangement but that there is a tile at each of the four positions neighboring $x$. Now, the tiles in the positions neighboring $x$ uniquely determine the missing tile at $x$, and since all pairs of adjacent tiles in $H$ come from adjacent positions in $[n]^2$, it follows that such a missing tile must be an isolated vertex of $H$. Once we add each such missing tile to the largest component of $H$ (by adding in the appropriately labeled directed edges), it follows from (C) that the largest component of $H$ contains each tile of $\mathcal{J}$ coming from the central $(n-2k-2) \times (n-2k-2)$ subgrid of $[n]^2$. Let $S_H$ denote the largest square subgrid contained in the largest connected component of $H$ at this juncture; we know from the above discussion that, with high probability, $S_H$ is a fully assembled $s \times s$ subgrid of $\mathcal{J}$ with $s \ge n-2k-2$.

We now finish the proof by showing that $\mathcal{J}$ has the following property with high probability: given any fully assembled $m \times m$ subgrid $M$ of $\mathcal{J}$ with $m \ge n-2k-2$, there is a unique way to assemble the tiles not in $M$ around $M$ to produce a feasible assembly of tiles on an $n \times n$ grid; of course, this final assembly of tiles must then coincide with $\mathcal{J}$.

Let us now describe an extension procedure that, with high probability, extends a given large fully assembled subgrid $M$ uniquely to $\mathcal{J}$, using the tiles not in $M$. This extension procedure will proceed by repeatedly extending $M$, first upward, then downward, then to the left, and, finally, to the right, adding an entire row or column of tiles at each step (thus ensuring that $M$ remains a subgrid at each stage). Suppose first that we wish to add a row of tiles to the top of $M$. Let $M'$ denote the set of tiles $t$ in the top row of $M$ not located at one of the two corners, and let $M''$ denote the set of two tiles at the top corners of $M$. For each $t \in M'$, we record all triples $(t', t'_l, t'_r)$ of tiles from the deck (not already in $M$) such that we may feasibly place $t'$ above $t$, $t'_l$ to the immediate left of $t'$, and $t'_r$ to the immediate right of $t'$. We then proceed as follows:

1. If no such feasible triple of tiles $(t', t'_l, t'_r)$ exists for some tile $t \in M'$, then we stop attempting to extend $M$ upward and change directions.
2. If there exist two distinct choices for the tile $t'$ over all recorded feasible triples $(t', t'_l, t'_r)$ for some tile $t \in M'$, then we abort.
3. If there exists a single choice for $t'$ (Though potentially more than one choice for $t'_l$ and $t'_r$) over all recorded feasible triples for each tile $t \in M'$, then we add a new row of tiles to the top of $M$ by first placing $t'$ above $t$ for each tile $t \in M'$. We then check if there exists a unique way to place two tiles (that are not already in $M$) feasibly above the two tiles in $M''$, and if so, we finish adding a new row to the top of $M$ by placing these two tiles in place; if we cannot find such a pair of tiles or if multiple choices exist for this pair, then we again abort.

Assuming that we have not aborted at any stage, we then continue to add rows to the top of $M$ until we are forced to change directions, and we then similarly extend $M$ downward, to the left, and, finally, to the right.

To bound the probability that this extension procedure fails to uniquely reconstruct $\mathcal{J}$ from some large fully assembled subgrid, we need to define two events. It will be convenient to first have some notation. Let $B_1 \subset [n]^2$ denote the set of vertices not contained in the central $(n-4k-4) \times (n-4k-4)$ subgrid of $[n]^2$, and let $B_2 \subset B_1$ denote the set of vertices in the four $(2k+2) \times (2k+2)$ subgrids at the four corners of $B_1$.

We first address the possibility of "failing in a corner" when extending a large subgrid. Let $E_1$ denote the event that there exists a pair $(u, v)$ with $u \in B_2$ and $v \in B_1$ such that some two edges incident to $u$ receive the same two colors under $\mathcal{J}$ as some two edges incident to $v$. We then have the following estimate.

**Claim 2.5.2** $\mathbb{P}(E_1) = o(1)$.

***Proof*** Let $Y_1$ denote the number of pairs $(u, v)$ which satisfy the conditions of the event $E_1$. It is easy to see that

$$\mathbb{E}[Y_1] = O(nk^3q^{-2} + k^2q^{-1}) = o(1);$$

the claim follows from Markov's inequality. □

Next, we address the possibility of "failing in the bulk of a row or column" when extending a large subgrid. Let $E_2$ denote the event that there exists a quadruple $(u, v, v', v'')$, where $u, v, v', v'' \in B_1$ and $u$ is not one of the four corners of $[n]^2$, such that either $v \neq u + e_1$ and the map $f\colon [-1,1] \times [0,1] \to [n]^2$ defined by $f(-1,0) = u + e_4$, $f(0,0) = u$, $f(1,0) = u + e_2$, $f(-1,1) = v'$, $f(0,1) = v$, and $f(1,1) = v''$ is feasible for $\mathcal{J}$, or the quadruple satisfies an analogous condition with respect to one of the three other directions. We then have the following estimate.

**Claim 2.5.3** $\mathbb{P}(E_2) = o(1)$.

***Proof*** Let $Y_2$ denote the number of quadruples $(u, v, v', v'')$ that satisfy the conditions of the event $E_b$. We may verify (after a somewhat tedious case analysis) that

$$\mathbb{E}[Y_2] = O((nk)^4 q^{-5} + (nk)^3 q^{-4} + (nk)^2 q^{-3}) = o(1);$$

the claim follows from Markov's inequality. □

If neither $E_1$ nor $E_2$ occurs, then it is easily seen by induction that our extension procedure extends any fully assembled $m \times m$ subgrid $M$ with $m \geq n - 2k - 2$ uniquely to $\mathcal{J}$, using the tiles not already in $M$. It follows that we may extend $S_H$ uniquely to $\mathcal{J}$ with high probability, proving the theorem. □

## 2.6 Conclusion

We conclude by reminding the reader of Conjecture 2.1.3, which asserts that the answer to the question of whether $\mathcal{J}(n,q)$ is reconstructible exhibits a sharp transition at $q \approx n/\sqrt{e}$. Here, we have established the 0-statement in Conjecture 2.1.3, using a simple counting argument. We have also proved the 1-statement in this conjecture for all $q \geq Cn$, where $C > 0$ is some absolute constant. As mentioned earlier, it is possible to use our methods to show that we may actually take $C$ as above to be any constant strictly greater than 1: roughly speaking, our estimates for the number of templates in the "small edge boundary" regime are very crude, and it is possible to do significantly better in this regime using stability results (see, [86]) for the isoperimetric inequality in $\mathbb{Z}^2$. However, showing that we may actually take $C$ as above to be any constant strictly greater than $1/\sqrt{e}$ appears to be completely out of the reach of our methods; we expect new ideas will be required to settle this problem.

Of course, one could also ask for the size of the window in the sharp transition predicted by Conjecture 2.1.3. By repeating the proof of the 0-statement of Theorem 2.1.3 with more careful estimates, we are led to the following refinement of Conjecture 2.1.3, whose 0-statement again follows from our counting argument.

**Conjecture 2.6.1** *Let $q = q(n) = n/\sqrt{e} + \log n + \alpha(n)$. As $n \to \infty$, we have*

$$\mathbb{P}(\mathcal{J}(n,q) \textit{ is reconstructible}) \to \begin{cases} 1 & \textit{if } \alpha(n) \to \infty, \textit{ and} \\ 0 & \textit{if } \alpha(n) \to -\infty. \end{cases}$$

Finally, it would be of interest to investigate higher-dimensional analogs of the problem considered here. For example, it would be interesting to decide if the analogous $d$-dimensional problem of reconstructing a random $q$-coloring of (the edges of) $[n]^d$ from its deck exhibits a sharp threshold at $q \approx n/e^{1/d}$ for each $d \geq 3$.

## ACKNOWLEDGMENTS

The first and second authors were partially supported by NSF grant DMS-1600742, and the second author also wishes to acknowledge support from EU MULTIPLEX grant 317532. Some of the research in this paper was carried out while the first author was visiting the Isaac Newton Institute for Mathematical Sciences at the University of Cambridge; the first author is grateful for the hospitality of the institute.

# 3

# Classifying Networks with *dk*-Series

**Marija Mitrović Dankulov[1], Guido Caldarelli[2], Santo Fortunato[3], and Dmitri Krioukov[4]**

[1]Scientific Computing Laboratory, Center for the Study of Complex Systems, Institute of Physics Belgrade, University of Belgrade, Pregrevica 118, 11080 Belgrade, Serbia
[2]IMT Alti Studi, Lucca, Italy
[3]Center for Complex Networks and Systems Research, School of Informatics and Computing and Indiana University Network Science Institute (IUNI), Indiana University, Bloomington, USA
[4]Department of Physics, Department of Mathematics, Department of Electrical & Computer Engineering, Northeastern University, Boston, MA, USA
MMD was supported by the Ministry of Education, Science, and Technological Development of the Republic of Serbia under project ON171017

## 3.1 Introduction

The theory of complex networks provides powerful tools for studying complex systems in various disciplines such as biology, social sciences, computer sciences, mathematics, and physics [210]. One of the main research directions in network science studies the structural properties of networks and how they affect the dynamical processes and functions of systems represented by these various networks [17]. The standard assumption is that a self-organizing system should evolve to a network structure that makes these dynamical processes, or network functions, efficient [209, 198, 50]. Thus, understanding the network structure also reveals the mechanisms underlying the evolution of the system represented by the network.

Topological structure of networks can be characterized by a great number of various measures describing the system organization at different levels. Measures such as degree, average neighbor degree, clustering coefficient, concentrations of small subgraphs, betweenness, the distribution of shortest paths, and spectral properties have been used to describe in quantitative manner features that are characteristic of wide classes of complex networks. It has been shown that many complex networks have fat-tailed degree distributions [16], possess the small world property [308], and are often organized in

Mitrović Dankulov, M., Caldarelli, G., Fortunato, S., and Krioukov, D., "Classifying Networks with *dk*-Series" in *Multiplex and Multilevel Networks*, edited by Battiston, S., Caldarelli, G., and Garas, A. © Oxford University Press 2019.
DOI: 10.1093/oso/9780198809456.003.0003

communities [96]. There is a common belief that the evolution of networks with similar structural properties is governed by the same mechanism. For instance, preferential attachment is often used for modeling networks with fat-tailed degree distributions [38]. Thus, classifying networks according to their topological properties is of great importance for identifying these mechanisms and better understanding the evolution, and indirectly the function and dynamics, of various complex systems.

However, classifying network via such graph measures is problematic, as there is no systematic way to determine which of them should be used. Besides, the measures are interdependent, that is, they positively or negatively correlate with each other in a complex way [303, 124, 289, 98, 63]. For these reasons, it is quite difficult to classify the structure of networks in a unique way using these topological measures and thus identify the evolution mechanisms characteristic for each class. For instance, the small world property has been found in many real networks, including social networks and interareal cortical networks in the primate brain. Yet in social networks, which are sparse graphs, this property is due to randomness in the linking patterns between the nodes, whereas in cortical networks it is trivially the consequence of network's high density. Therefore, the evolution of these two networks has been driven by different mechanisms, so that they cannot be assigned to the same network class.

One way to address the problem of interdependence among network properties is to find which of them are significant for a given network, and thus for its function. The standard procedure for the identification of a significant property $X$ and its dependence on some other property $Y$ is to generate a set of random graphs that have property $Y$ but are random in all other respects, and then to check whether the property $X$ is also characteristic of these graphs. If this is the case, then obviously property $X$ is not interesting and relevant for the network function and dynamical processes running on it. We conclude that property $X$ is a statistical consequence of property $Y$, and $Y$ fully describes the structure of the network. Mechanisms generating network property $Y$ can be thus considered to be relevant for the network evolution and dynamics. If $X$ is not a typical property of these random graphs, one cannot conclude anything about the relevance of property $X$. The only conclusion that would follow from this is that property $X$ is independent of property $Y$, but that does not mean that it is also independent of some other network property.

The identification of significant network features using the procedure described above raises another, equally important, question about the choice of null models. Since there are infinitely many network properties $Y$, there are infinitively many null models defined by property $Y$, and these can be used to test the statistical significance of any other property $X$ [297]. For example, for most properties $X$, including motifs [195], their significance is tested with respect to random graphs with the same degree distribution. Although the choice of degree distribution as a $Y$ property can seem natural, given the fundamental role played by it [210], there is no evidence that this choice of null model is less arbitrary than others. In general, there can be some other property which can be explanatory for both $Y$ (here, the degree distribution) and $X$. Thus, one needs to identify the right reference property or properties $Y$ in the null model that should be used for the testing of the (statistical) significance of property $X$.

In a recent work, Orsini *et al.* [219] proposed a way to identify such basic properties, enabling us to do a complete systematic description and unique classification of the structure of real networks. It is based on a set of properties known as the *dk-series* [182], a converging series of basic interdependent degree- and subgraph-based properties that characterize the local network structure at an increasing level of detail. It has been shown [182, 219] that the *dk*-series also defines a corresponding series of null models or random graph ensembles. These random graph models have exactly the same distribution of subgraphs of size $d$ for all $d$-ples of nodes with degree $(k_1, k_2, k_3, \ldots, k_d)$ as in the real network. Or, to be precise, they are random graphs with fixed average degree, degree distribution, degree correlations, clustering, and so on. In Ref. [219], the authors used this methodology to quantify the randomness of six real single-layer networks, of very different function and dynamics. They showed that random graphs with fixed degree distribution, degree–degree correlations, average clustering, and degree-dependent average clustering reproduce all relevant topological properties for most networks. Here, we apply this approach to three networks and show that they differ in the randomness of their structure. We show that although many network properties can be reduced to specific degree- and subgraph-based characteristics, some of them cannot be explained with *dk*-series.

In recent years, a lot of attention in network science has been devoted to networks in which the same set of nodes are connected with multiple links of different types. These networks are referred to as multiplex or multilayer networks, since they consist of correlated single-layer networks composed of links of the same type. Many of the topological measures used to describe the structure of single-layer networks have been adapted in order to characterize the structure of layers and correlations between them [37]. We show how *dk*-series can be extended to describe in a systematic way the structure of multiplex networks using *dk*-annotated series [79].

## 3.2 *dk*-Series for single-layer networks

As indicated in the previous section, one needs to find an ordered set of reference properties of networks $Y_0, Y_1, \ldots$, satisfying some criteria. The first criterion is *inclusiveness*: every subsequent property provides more details about the network structure than its predecessor. Formally, this is equivalent to the requirement that networks with property $Y_d$, $d > 0$, should also have all properties prior to it, that is, all properties $Y_{d'}$, where $0 \leq d' < d$. The second criterion is *convergence*, that is, the minimal set of properties has to be finite, that is, the last property in series $Y_D$ should fully characterize the adjacency matrix of any given graph. The $Y$-series that satisfies these conditions allows us to claim that, for any property $X$ that is deemed important in a given real network, we can find a minimal $d^*$ such that the property $Y_{d^*}$ explains property $X$. The convergence of the series ensures the existence of some $d^*$, while the inclusiveness means that random networks with $Y_d$ $(d = (d^* + 1), \ldots, D)$ also have property $X$, so if we go to higher values of $d$, the random network has property $X$ along with other significant properties. This enables the classification of network structure in a systematic manner. Several approaches, including motifs [195], graphlets [314], and similar constructions [212], try to fully characterize the structure of networks by using relatively small set of properties, but they all violate the inclusiveness condition. On the other hand, one can still define many $Y$-series satisfying both conditions. We chose *dk-series* [182], the most natural choice due to their simplicity and the fact that they are a combination of subgraph- and degree-based characteristics of networks.

In these series, properties $Y_d$ are *dk*-distributions. Each *dk*-distribution is actually a collection of distributions, stating how the subgraphs of size $d$ are distributed over nodes with degrees $k, k', k'', \ldots, k^d$ in a graph $G$. Note here that the isomorphic subgraphs of $G$ involving nodes of different degrees are thus counted separately. Specifically, the $0k$ "distribution" is simply the average degree $\bar{k} = \frac{2M}{N}$, where $N$ and $M$ are the number of nodes and links in a given graph, while the $1k$ distribution is the number of subgraphs of size 1, nodes, with the degree $k$, that is, the standard degree distribution

$$P(k) = \frac{N(k)}{N}, \tag{3.1}$$

where $N(k)$ is the number of nodes of degree $k$. The $2k$ distribution counts how many nodes of degrees $k$ and $k'$ are forming subgraphs with two nodes, and is known as joint degree matrix $P(k, k')$

$$P(k, k') = \frac{\mu(k, k')M(k, k')}{2M}, \tag{3.2}$$

where

$$\mu(k, k') = \begin{cases} 2 & \text{if } k = k', \\ 1 & \text{otherwise.} \end{cases} \tag{3.3}$$

The $3k$ distribution is a set of two distributions corresponding to two non-isomorphic subgraphs of size 3: wedges $\wedge$ and triangles $\Delta$. It characterizes the connectivity patterns between triples of nodes of degrees $k$, $k'$, and $k''$:

$$P_\wedge(k', k, k'') = \mu(k', k'')\frac{N_\wedge(k', k, k'')}{2W}, \tag{3.4}$$

$$P_\Delta(k, k', k'') = \nu(k, k', k'')\frac{N_\Delta(k, k', k'')}{6T}, \tag{3.5}$$

where $W$ and $T$ are the total numbers of wedges and triangles in the network, and

$$\nu(k, k', k'') = \begin{cases} 6 & \text{if } k = k' = k'', \\ 1 & \text{if } k \neq k' \neq k'', \\ 2 & \text{otherwise,} \end{cases} \tag{3.6}$$

so that both $P_\wedge(k',k,k'')$ and $P_\triangle(k,k',k'')$ are normalized, and $\sum_{k,k',k''} P_\wedge(k',k,k'') = \sum_{k,k',k''} P_\triangle(k,k',k'') = 1$. We could continue in a similar manner to obtain a $4k$ distribution that consists of six distributions, each corresponding to one of the six non-isomorphic subgraphs of size 4 and so on until we reach $d = N$, the $Nk$-distribution, which characterizes the whole adjacency matrix of a given graph (see Figure 3.1).

The *dk*-series is directly related to some of the standard topological measures from complex network theory. Besides the already mentioned average degree and degree distribution, which are directly related to the $0k$ and $1k$ distributions, the $2k$ distribution defines the node degree correlations in networks, or network's assortativity. The average neighbor degree $\overline{k}_n n(k)$ is a projection of $P(k,k')$ via

$$\overline{k}_{nn}(k) = \frac{\sum_{k'} k' P(k,k')}{\sum_{k'} P(k,k')}.$$

The average clustering coefficient $\overline{c}$ and degree-dependent clustering coefficient $\overline{c}(k)$ can be calculated based on the number of triangles in the network. Specifically,

$$\overline{c} = \frac{1}{N} \sum_i \frac{2\Delta_i}{k_i(k_i - 1)},$$

and

$$\overline{c}(k) = \frac{6T}{N} \frac{\sum_{k',k''} P_\triangle(k,k',k'')}{k(k-1)P(k)}, \tag{3.7}$$

where $\Delta_i$ is the number of triangles composed of node $i$ and its neighbors, while $T$ is the total number of triangles. In general, the arbitrary *dk*-distribution characterizes both degree correlations between nodes at the hop distances $d' < d$, and the frequencies of $d'$-sized subgraphs, $d' \leq d$, in graph $G$.

One can easily see that *dk*-series is inclusive. The $(d+1)k$-distribution contains the same information about the network structure at the level $d$ as the *dk*-distribution, plus some additional information about the degree correlations at the level $d+1$. Specifically, the $1k$-distribution defines the average degree ($0k$-distribution), via

$$\overline{k} = \sum_{k'} k' P(k'),$$

while the $1k$-distribution can be obtained from the $2k$ distribution as

$$P(k) = \frac{\overline{k}}{k} \sum_{k'} P(k,k').$$

Similarly, the $3k$-distribution defines the $2k$-distribution by

$$P(k,k') = \frac{1}{k+k'-2}\sum_{k''}\left\{\frac{6T}{M}P_{\triangle}(k,k',k'') + \frac{W}{M}\left[P_{\wedge}(k',k,k'') + P_{\wedge}(k,k',k'')\right]\right\}. \tag{3.8}$$

The opposite does not hold, that is, knowing the $dk$-distribution will not allow one to infer anything about the $(d+1)k$-distribution, meaning that higher values of $d$ correspond to a greater level of details about the network structure.

The number of non-isomorphic subgraphs, and thus the number of distributions needed to characterize network topology at the level $d$, grows exponentially with $d$; hence, the calculation of $dk$-series becomes a computationally intensive task for higher values of $d$. One could argue that just counting the number of $d$-sized subgraphs in a given network regardless of their node degrees should be enough for the description of network structure in a systematic manner [195, 314, 212]. The subgraph-based series obtained from the count of $d$-subgraphs without including information about the degree, which we can call a $d$-series, satisfies the convergence condition, and the statistics for $d = N$ subgraphs would also fully describe the topology of a given network, but, unlike $dk$-series, they are not inclusive. Careful analysis of the first four elements of a $d$-series clearly demonstrates its non-inclusiveness. For the $d$-series, the zeroth element is not defined, while the number of nodes $N$ and number of edges $M$ are the properties corresponding to $d = 1$ and $d = 2$, respectively. These two quantities are independent of each other, that is, knowing the number of edges does not allow one to tell much about the number of nodes in a network. Similarly, the properties at the level $d = 3$, the number of triangles $T$ and wedges $W$, define neither the size nor the density of the network [219]. This analysis demonstrates that the elements in the $d$-series are independent of each other and that each of them conveys a different kind of information about network topology.

Figure 3.1(b) illustrates the inclusiveness and convergence of $dk$-series, and also suggests that all graphs with $N$ nodes and $M$ edges constitute a set of random graphs $\mathcal{G}_{0k} = \mathcal{G}_{N,M}$ with the same $0k$ property. Graphs with the same degree sequence, the $1k$ property, form a smaller set of graphs $\mathcal{G}_{1k}$, which is a subset of $\mathcal{G}_{0k}$, and so on. Each set of graphs with a given $dk$-distribution, known as *dk-graphs*, is at the same time a subset of $(d-1)k$- and a superset of $(d+1)k$-graphs. It follows from this that a sequence of $dk$-distributions defines a sequence of random graph ensembles (null models). In order to compare a real network with the random graphs which have the same $dk$-properties, and thus quantify the randomness of its structure of the real network, one needs a maximum entropy ensemble of these graphs or *dk-random graphs* [182]. All graphs in $dk$-random graphs have equal sampling probability $P(G) = 1/\mathcal{N}_d$, where $\mathcal{N}_d$ is the number of $dk$-graphs. Each collection of $dk$-distributions is more informative about the network structure and thus more constraining than $(d-1)k$-distributions, that is, $\mathcal{N}_0 \geq \mathcal{N}_1 \geq \cdots \geq \mathcal{N}_N = 1$. The size of the final ensemble at level $N$ is clearly equal to 1, since it just contains the network with the exact adjacency matrix. The number $\mathcal{N}_d$ is too

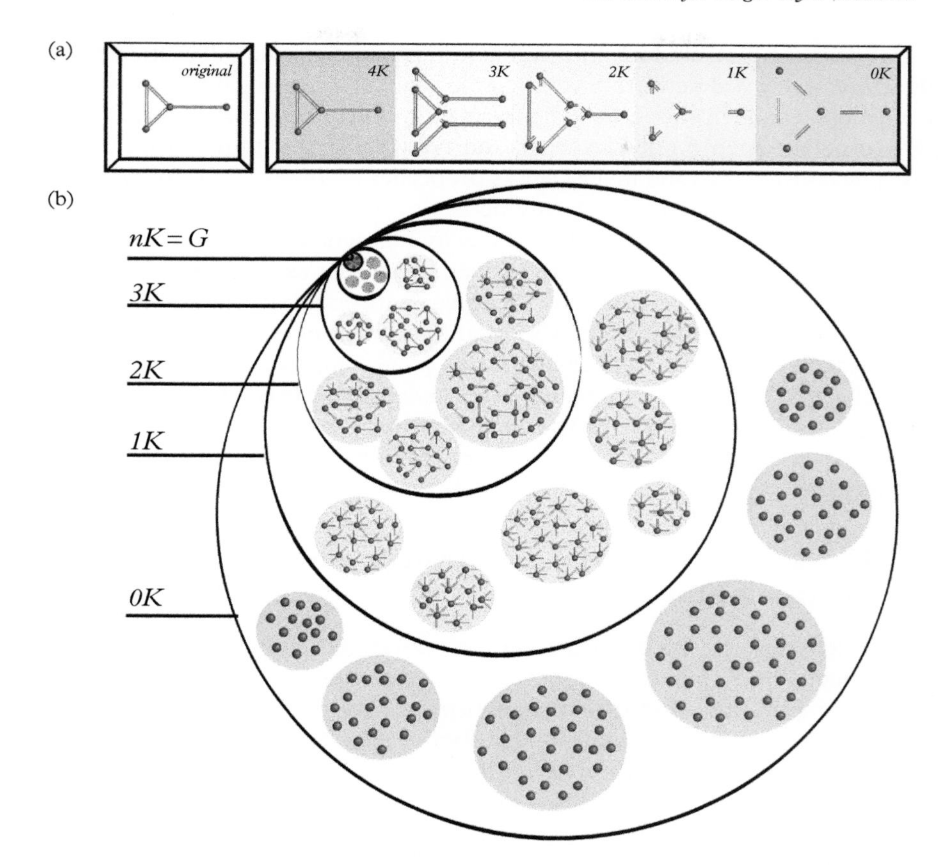


**Figure 3.1** *The illustration of* dk-*series. (a) The* dk-*distributions for a graph of size 4. The* 4k-*distribution is the graph itself. The* 3k-*distribution consists of its three subgraphs of size 3: one triangle connecting nodes of degrees 2, 2, and 3, and two wedges connecting nodes of degrees 2, 3, and 1. The* 2k-*distribution is the joint degree distribution in the graph. It specifies the number of links (subgraphs of size 2) connecting nodes of different degrees: one link connects nodes of degrees 2 and 2, two links connect nodes of degrees 2 and 3, and one link connects nodes of degree 3 and 1. The* 1k-*distribution is the degree distribution in the graph. It lists the number of nodes (subgraphs of size 1) of different degree: one node of degree 1, two nodes of degree 2, and one node of degree 3. The* 0k-*distribution is just the average degree in the graph, which is 2. (b) The inclusiveness and convergence of the dk-series are illustrated via the hierarchy of* dk-*graphs, which are graphs having the same* dk-*distribution of a given graph* G *of size* N. *From Ref. [219].*

large, especially for small values of $d$ (exact or approximate calculations for $d = 0, 1, 2$ can be found in [32, 20]), making the construction of the whole set of *dk*-graphs impossible. Thus, one needs to sample *dk*-random graphs uniformly in order to be able to compare them with a given real network and properly test the significance of different topological properties.

The *dk*-series and *dk*-random graphs enable the systematic and full characterization of the structure of any real network by finding the value of $d$ for which all $d'k$-distributions for $d' > d$ do not contain any additional information of the network structure. This means that any topology metric one can define on network $G$ is captured with *dk*-random graphs. The convergence and inclusiveness properties of *dk*-series ensure that this value $d$ exists, that is, they guarantee that any network property $X$ of any given network $G$ can be reproduced with any desired accuracy by high-enough $d$. Clearly, all properties are reproduced exactly for $d = N$, but the question is whether there is a value $d < N$ for which all relevant topological properties of a given network are captured with *dk*-random graphs. By finding this value $d$, one also quantifies the randomness of the structure of a given network. The entropy of *dk*-ensembles is $S_d = \ln \mathcal{N}_d$, and it is a nonincreasing function of $d$, that is, the *dk*-random graphs are *less random and more structured*, the higher $d$ is. In the following section, we demonstrate how one can classify single-layer networks based on their *dk*-randomness by applying the procedure described in Ref. [219] to three real networks.

### 3.2.1 Classifying single-layer networks based on their *dk*-randomness

First, we briefly discuss the constructibility of *dk*-random graphs, and the problem of sampling graphs uniformly at random from the sets of *dk*-graphs. Here, we emphasize that, in *dk*-graphs, the *dk*-distribution constraints are sharp, that is, all graphs in *dk*-graphs set have exactly the same *dk*-distribution. Given a real network $G$, there exist two ways to sample *dk*-random graphs: *dk*-randomize $G$, generalizing the randomization algorithms in Refs. [186, 187], or construct random graphs with $G$'s *dk*-sequence from scratch [182, 117], also called direct construction [148, 76, 147, 22]. We chose the first option, *dk*-randomization, due to its simplicity and the existence of algorithms that enable the uniform sampling of *dk*-random graphs for values of $d$ greater than 2 (see the detailed discussions about construction algorithms in [182, 219]).

The *dk*-randomization is an edge-swapping procedure where pairs of edges are swapped at random, starting from $G$, such that the *dk*-distribution is preserved at each swap. Figure 3.2 illustrates permitted swaps of edges for each *dk*-distribution. Specifically, to preserve $0k$-distribution, average degree, we disconnect a pair of nodes and connect two other, non-neighboring, nodes. The graphs obtained in this procedure are Erdős-Rényi graphs $\mathcal{G}_{N,M}$ of fixed size $N$ and average degree $2M/N$. To preserve the degree sequence ($1k$-distribution), we chose at random a pair of edges and swapped their targeting nodes, while, for the $2k$-distribution, we swapped edge pairs only if there were at least two nodes of equal degrees adjacent to different edges belonging to this pair. Allowed $3k$-swaps are then $2k$-swaps that preserve $3k$-distribution, the same connectivity patterns between the triplets of nodes with respect to node degrees. From this and the inclusiveness of *dk*-series, it follows that $(d+1)k$-swaps form a subset of *dk*-swaps for $d > 0$ [182]. During all these rewiring procedures, the edge swapping is only allowed if it does not lead to the creation of multiple edges between the same pair of nodes.

There are many concerns regarding the described rewiring procedure [316], two of which are particularly important: (1) the ergodicity of the rewiring process, that is, whether any two pairs of graphs with the same *dk*-properties are connected with a chain of *dk*-swaps; (2) the uniformity of the rewiring process, that is, how close to uniform sampling the *dk*-swap Markov chain is after its mixing time is reached. It has been shown that *dk*-swapping processes for values of $d = 1, 2$ are ergodic [186, 187, 67], while it is common belief that there is no ergodic edge swapping, of any type, that preserves the 3*k*-distribution, and thus *dk*-distributions for values $d \geq 4$, although a rigorous proof of this is lacking at the moment [219]. When it comes to the uniformity of the *dk*-swapping process for $d = 0, 1, 2$, it has been shown that if the edge-swap process is done correctly, then the sampling is uniform [2, 11].

Since the 2*k*-random graphs do not capture all topological properties for most of the tested real networks [137, 219], we need algorithms that will allow us to go beyond preserving only 2*k*-properties. The *dk*-targeting $d'k$-preserving rewiring, where $d' < d$, has proven to be a good choice for generating random graphs with the same *dk*-properties as in the considered real network [182, 219]. This procedure incorporates the following modification of the $d'k$-rewiring algorithm: the $d'k$-swap is accepted with probability $\min(1, \exp(-\beta \Delta H))$, where $\beta$ is the inverse temperature of this simulated annealing process, and $\Delta H$ is the change in the $L^1$ distance between the *dk*-distribution in the current graph and the targeted *dk*-distribution before and after the swap. The numerical experiments with 3*k*-targeting rewiring have shown that this process does not converge for most real networks [219], due to the extremely constraining nature of the 3*k*-distribution. Therefore, it is reasonable to retreat to numeric investigations of 2*k*-random graphs in which, in addition to the 2*k*-distribution, some substatistics of the 3*k*-distribution are fixed. In particular, we consider 2.1*k*-random graphs, which have the same 2*k*-distribution and value of the average clustering coefficient $\bar{c}$ as the given real network, and 2.5*k*-random graphs with the same 2*k*-properties and average clustering coefficient $\bar{c}(k)$ of nodes of degree $k$ [117]. Since 2.1*k*- and 2.5*k*-statistics are fully defined by the 3*k*-distribution, and 2.1*k* is defined by 2.5*k*, the 3*k*-random graphs comprise a subset of 2.5*k*-random graphs, which are, in turn, subsets of 2.1*k*-random graphs, that is, $\mathcal{N}_2 > \mathcal{N}_{2.1} > \mathcal{N}_{2.5} > \mathcal{N}_3$. As a consequence, if a certain topological property of real networks is captured by 2.5*k*-random graphs, it will be also captured by 3*k*-random graphs, while the opposite is not generally true.

The scheme of algorithm(s) that we use for creating a set of *dk*-random graphs for $d = 0, 1, 2, 2.1, 2.5$ is given in Figure 3.2, while their detailed description and a link to a Web page with publicly available software can be found in Ref. [75]. The *dk*-random graphs for $d = 0, 1, 2$ are created using the standard *dk*-swapping described above. Although it is known that these procedures for general graphs do not lead to a uniform sampling of *dk*-random graphs, unlike their modified versions [2, 11], it has been shown that, for power-law distributions, the obtained sample of uniform graphs is very close to uniform. To generate *dk*-random graphs for $d = 2.1, 2.5$, we start with a 2*k*-random graph and apply to it a described 2*k*-preserving 2.*xk*-targeting ($x = 1, 5$) rewiring process (see Figure 3.2). For this, we use a modified version of the algorithm [63, 219], which ensures the convergence for all networks.

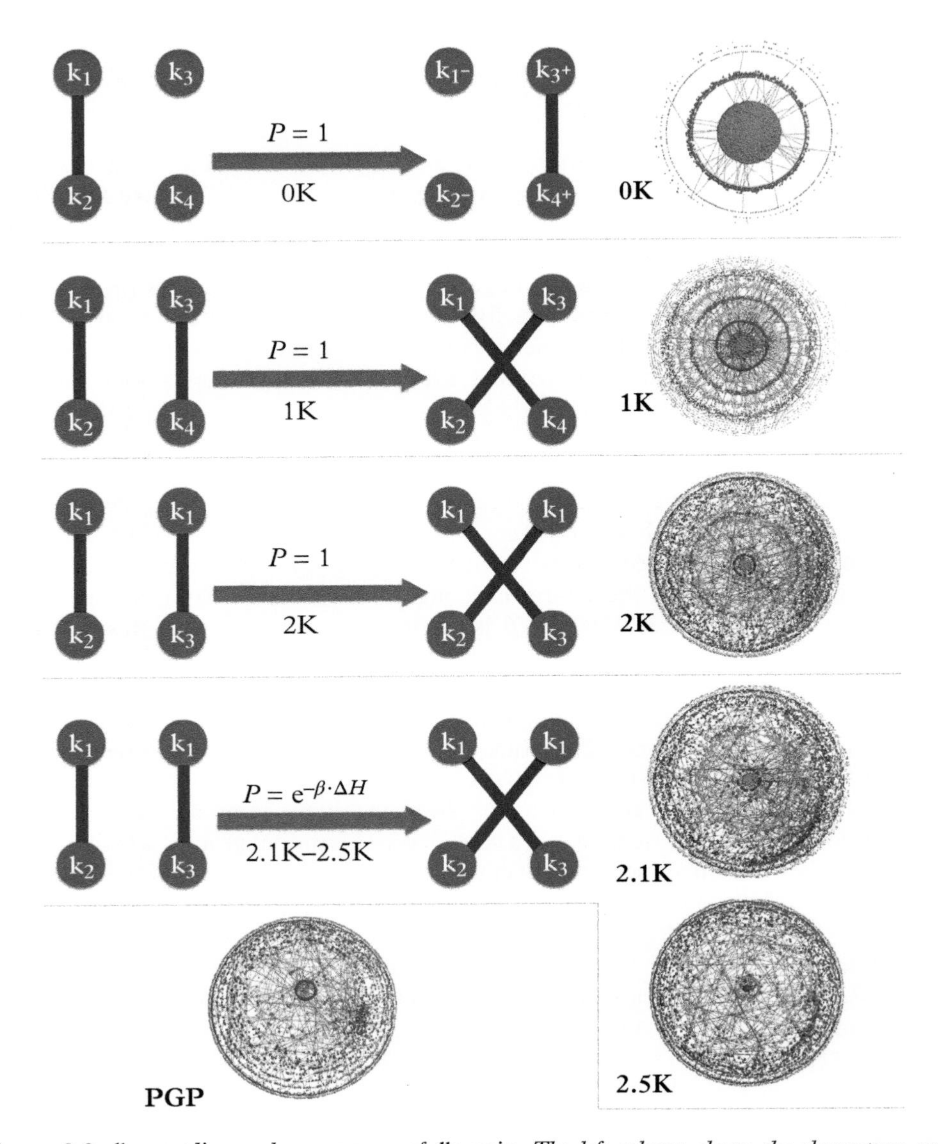


**Figure 3.2** dk-*sampling and convergence of* dk-*series. The left column shows the elementary swaps of* dk-*randomizing (for* d $= 0, 1, 2$*) and* dk-*targeting (for* d $= 2.1, 2.5$*) rewiring. The nodes are labeled by their degrees, and the arrows are labeled by the rewiring acceptance probability. In* dk-*randomizing rewiring, random (pairs of) edges are rewired preserving the graph's* dk-*distribution (and consequently its* d′K-*distributions for all* d′ < d*). In 2.1*k- *and 2.5*k-*targeting rewiring, the moves preserve the* 2k-*distribution, but each move is accepted with probability* p *designed to drive the graph closer to a target value of average clustering* $\bar{c}$ *(2.1*k*) or degree-dependent clustering* c(k) *(*2.5k*):* $\mathrm{p} = \min(1, e^{-\beta \Delta H})$, *where* $\beta$ *is the inverse temperature of this simulated annealing process,* $\Delta H = H_a - H_b$, *and* $\mathrm{H}_{\mathrm{a,b}}$ *are the distances, after and before the move, between the current and target values of clustering:* $\mathrm{H}_{2.1\mathrm{k}} = |\bar{c}_{current} - \bar{c}_{target}|$ *and* $\mathrm{H}_{2.5\mathrm{k}} = \sum_i |\bar{c}_{current}[\mathrm{k_i}] - \bar{c}_{target}[\mathrm{k_i}]|$. *The right column shows LaNet-vi [29] visualizations of the results of these dk-rewiring processes, applied to the Pretty Good Privacy (PGP) network, visualized at the bottom of the left column. The node sizes are proportional to the logarithm of their degrees, while the color reflects node coreness [29]. As* d *grows, the shown* dk-*random graphs quickly become more similar to the real PGP network. From Ref. [219].*

To quantify the randomness of the real network, that is, to determine the value of $d$ for which $dk$-random graphs capture most of its topological properties, we adopt the following procedure. For a given real network, we calculate its average degree, degree distribution, degree correlations, average clustering coefficient, and averaging clustering coefficient of nodes of degree $k$; then, based on this, we generate 20 $dk$-random graphs, using the methodology described in the previous paragraph, for $d = 0, 1, 2, 2.1, 2.5$. Then, for each sample, we compute a variety of network properties and compare their values with the corresponding ones obtained for the real network. The value of $d$ for which the considered properties of $dk$-random graphs are in reasonable agreement with the ones of a real network determines the randomness of its structure. The higher the value of $d$, the more structured and less random a given network is.

In Ref. [219] the authors performed an extensive set of numeric experiments with six real, very different networks with respect to their function. Here, we demonstrate the described procedure by applying it to three of these networks: the Internet at the level of autonomous systems (INTERNET) [183], a technosocial web of trust among users of the distributed Pretty Good Privacy (PGP) cryptosystem [39], and a functional MRI (fMRI) map of the human brain (BRAIN) [85]. In the first network, INTERNET, the nodes are so-called autonomous systems (ASs; organizations owning parts of the Internet infrastructure), and there is a link between two ASs if they have a business relationship in which they exchange Internet traffic. The nodes in the second network we consider, PGP, are users' PGP certificates, while the edges denote the existence of trust between two users. We consider here only the largest connected component of the PGP network. The third network considered, BRAIN, is the largest component of an fMRI map of the human brain, where voxels (representing small areas of a resting brain, approximately 36 mm$^3$ in volume) are represented with nodes, and an edge exists between two voxels if the correlation coefficient of the fMRI activity of the voxels exceeds 0.7. We chose these three networks because they have different values of $d$ for which $dk$-random graphs capture their structural properties [219]. In particular, most of the considered properties for INTERNET are reproduced with $2k$-random graphs, which makes it the most random network among the three networks, while, to reproduce the same properties of the PGP network, we need graphs with preserved $2.5k$-distribution. Some of the properties of the BRAIN network are not reproduced even with $2.5k$-random graphs, meaning that this network is the least random one among these three networks (Figures 3.3–3.7 and Tables 3.1 and 3.2).

The properties that we use to compare the structure of real networks with the ones of random graphs can be divided into three categories: microscopic, mesoscopic, and macroscopic. The microscopic properties describe the networks' structure at the level of individual nodes and subgraphs of small sizes (see Figures 3.3 and 3.4). Some of these properties, namely, average degree, degree distribution, average degree of nearest neighbors, average clustering coefficient, and average clustering coefficient of nodes of degree $k$, are fixed by the corresponding $dk$-distributions. On the other hand, the concentration of subgraphs of size 3 and 4 [219], as well as the distribution of the number of common neighbors shared by a pair of nodes, are not fixed by $dk$-distributions for $d < 3$. The distribution of common neighbors equals the probability that two connected

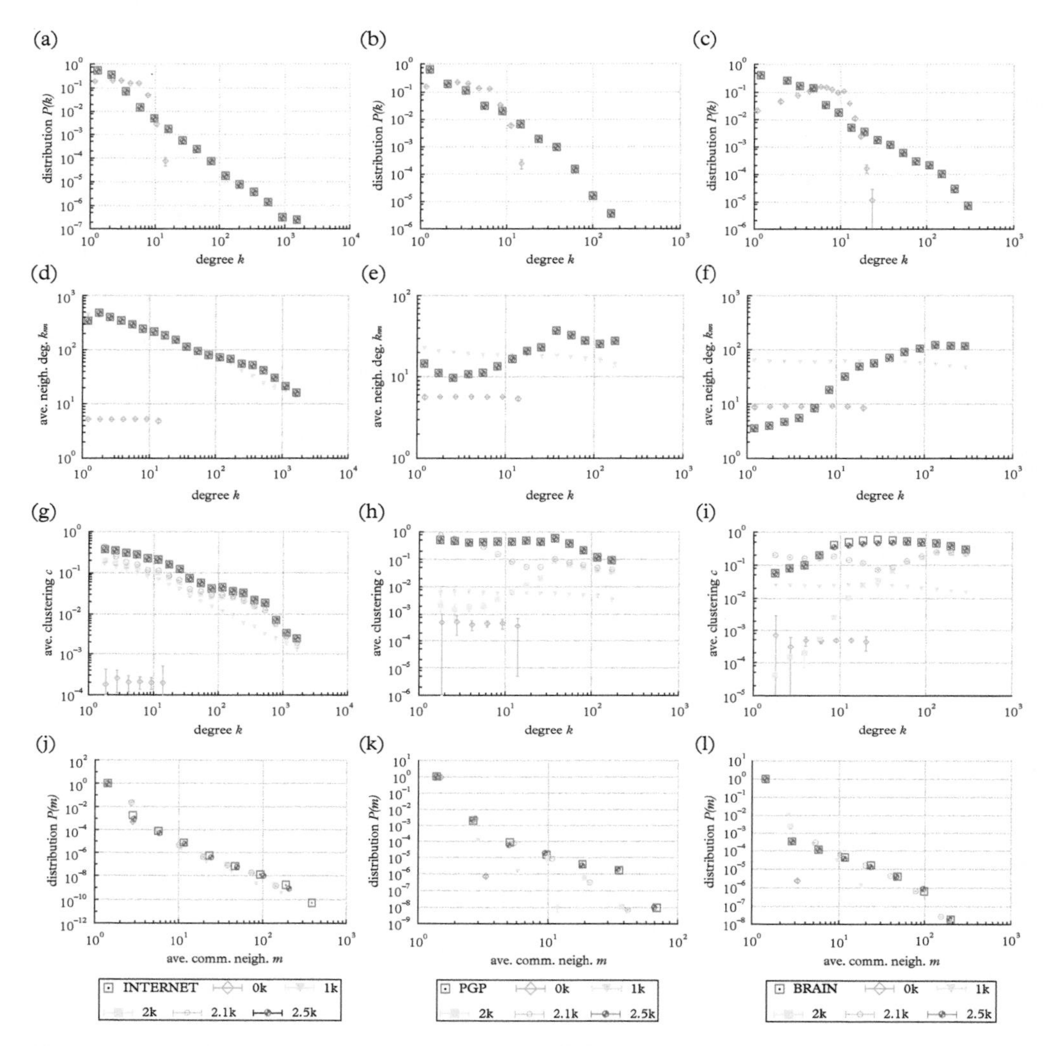


**Figure 3.3** *Microscopic properties of real complex networks and their* dk-*random graphs. The first nine panels show topological properties fixed by* dk-*distributions: the degree distribution* P(k) *for (a) INTERNET, (b) PGP, and (c) BRAIN; the average degree of nearest neighbors* $\bar{k}_{nn}(k)$ *for (d) INTERNET, (e) PGP, and (f) BRAIN; and the degree-dependent average clustering coefficient* $\bar{c}(k)$ *for (g) INTERNET, (h) PGP, and (i) BRAIN. The last three panels show the distribution of the number of common neighbors* P(m)*, which is not fixed by* dk-*distributions for* $d < 3$*: (j) INTERNET, (k) PGP, and (l) BRAIN. Adapted from Ref. [219].*

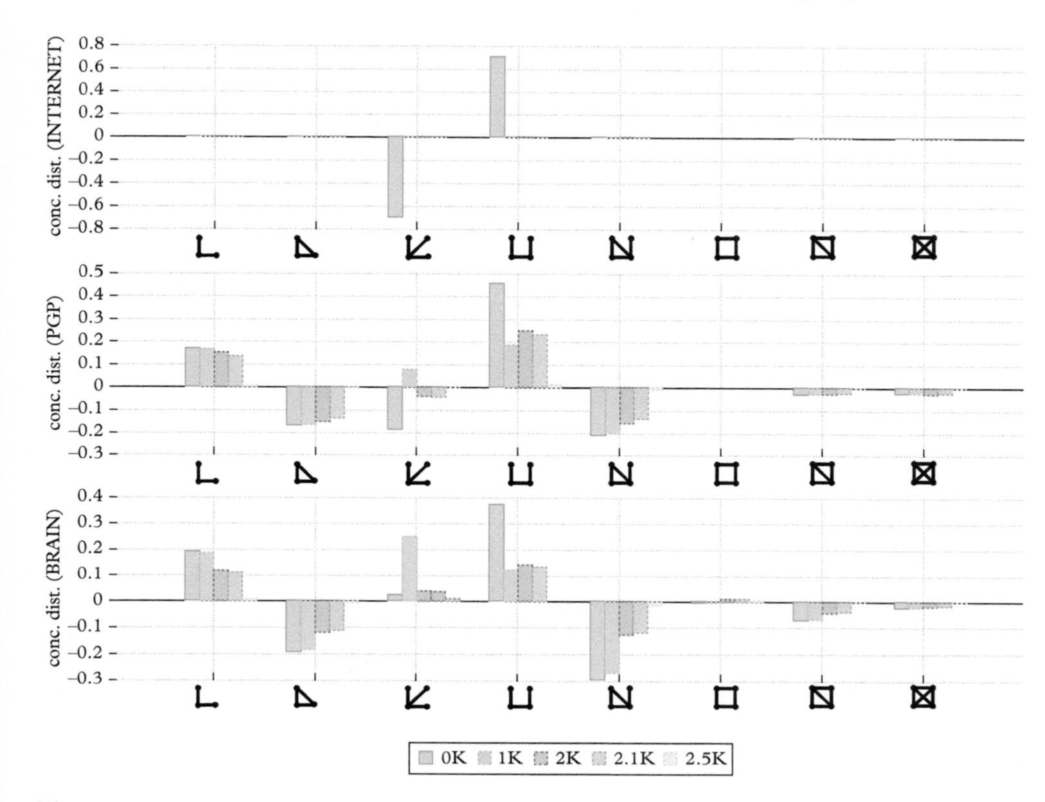


**Figure 3.4** *Density of subgraphs of sizes 3 and 4. The average absolute difference between the subgraph concentration in the* dk*-random graphs and its concentration in the real networks: INTERNET (top), PGP (middle), and BRAIN (bottom). Adapted from Ref. [219].*

nodes have $m$ common neighbors and is exactly fixed by the $3k$-distribution. Mesoscopic properties depend on both local and global network organization. Here, we consider $k$-coreness [9] and $k$-density [250] (see Figure 3.5). A node has $k$-coreness equal to $k$ if it belongs to $k$-core of the original graph, which is the largest induced subgraph of graph in which every node has degree at least $k$. Similarly, an edge has $k$-denseness equal to $k$ if it belongs to the largest induced subgraph of the original graph in which all edges have multiplicity at least $k$. Macroscopic properties are truly global: betweenness, the distribution of hop lengths of shortest paths, and spectral properties (see Figure 3.6 and Tables 3.1 and 3.2). We measure the distance between real and $dk$-random graphs with Kolmogorov–Smirnov distances between the distributions of all the considered properties (see Figure 3.7).

For all three networks and for most of the considered properties, we observe a nice convergence as $d$ increases, that is, there is no statistically significant difference between the property in the real network and in its $2.5k$-random graphs. Although this is expected

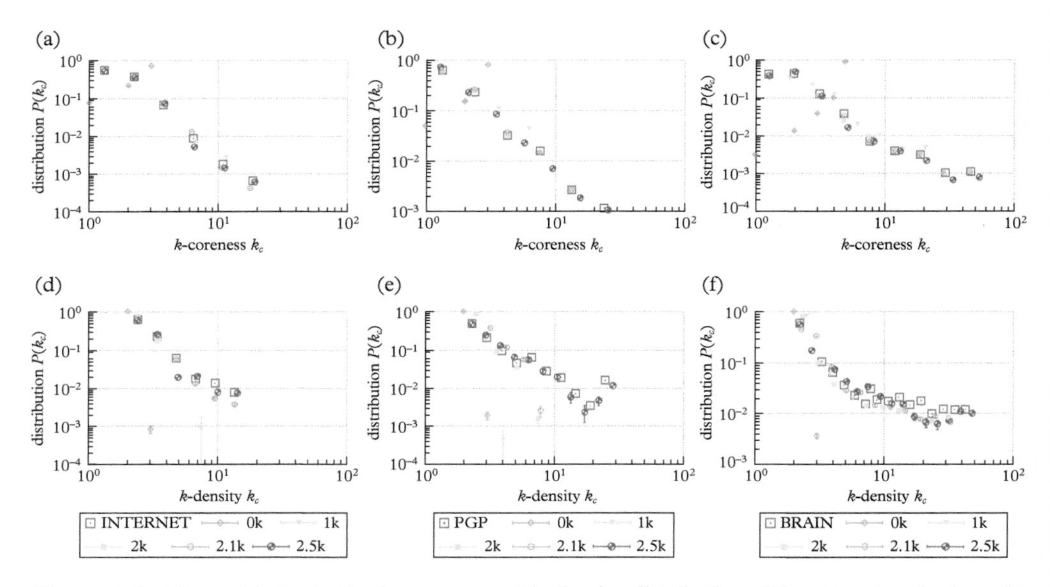


**Figure 3.5** *Mesoscopic properties:* k*-coreness and* k*-density distributions. The upper panels show the* k*-coreness of (a) INTERNET, (b) PGP, and (c) BRAIN networks, and the lower panels show their* k*-density: (e) INTERNET, (f) PGP, and (g) BRAIN. The* $k_c$*-core of a graph G is the maximal subgraph of G in which all nodes have degree at least* $k_c$. *A node has* k*-coreness* $k_c$ *if it belongs to the* $k_c$*-core but not to the* $(k_c + 1)$*-core. The* $k_c$*-dense subgraph is the maximal subgraph of a graph in which all edges have multiplicity* $(k_c - 2)$*; the multiplicity of an edge is the number of triangles the edge is part of. Adapted from Ref. [219].*

for microscopic properties that are fixed with *dk*-distributions, there is no reason to expect convergence in the case of small subgraph frequencies, in the distribution of the number of common neighbors, or for mesoscopic or macroscopic properties. Figure 3.4 shows that the relative difference between subgraph frequencies in real and 2.5*k*-random graphs is very close to zero for subgraphs of sizes 3 and 4. For INTERNET, this property is already captured with 1*k*-graphs, while it is clear that, for the BRAIN and PGP networks, one needs to fix the degree-dependent clustering coefficient in order to observe the same motif count as in the real systems. Mesoscopic properties are reproduced with 2.5*k*-graphs for the BRAIN and PGP networks, and with 2*k*-random graphs for the INTERNET network (see Figure 3.5). While betweenness and average shortest-path distance require 1*k*- and 2.5*k*-random graphs for INTERNET and PGP, respectively, such properties are not captured even with 2.5*k* graphs for BRAIN (see Figure 3.6). Table 3.1 shows that the largest eigenvalue of the adjacency matrix is closely, but not exactly, reproduced by $d = 2.5$ for all three networks. The spectral gap, the difference between the largest and second largest eigenvalue of the adjacency matrix, given in Table 3.2, shows that 2*k*- and 2.1*k*-random graphs are better connected and interlinked, compared to real networks. Figure 3.7 shows that the Kolmogorov–Smirnov distances calculated for the distributions of INTERNET and PGP are either zero or

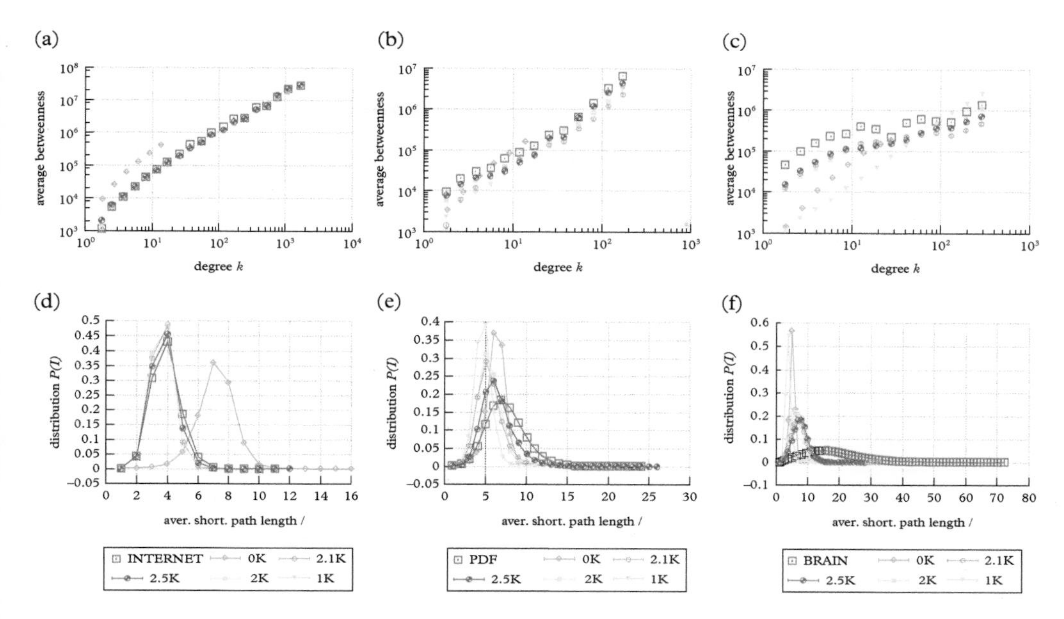


**Figure 3.6** *Macroscopic properties: betweenness and average shortest hop distance. The average betweenness* $\bar{b}(k)$ *of nodes of degree* k *is shown in the upper panels: (a) INTERNET, (b) PGP, and (c) BRAIN. The lower panels show the distribution* P(l) *of the length* l *of the shortest paths between all pairs of nodes: (d) INTERNET, (e) PGP, and (f) BRAIN. Adapted from Ref. [219].*

**Table 3.1** *Largest eigenvalues of the adjacency matrix for the three networks considered and their corresponding dk-graphs. For the latter, we show averages across different realizations for each d, and their standard deviations in parentheses.*

| | Original | 0*k* | 1*k* | 2*k* | 2.1*k* | 2.5*k* |
|---|---|---|---|---|---|---|
| **INTERNET** | 67.17 | 5.36 (0.01) | 56.02 (0.33) | 61.15 (0.03) | 61.32 (0.06) | 65.34 (0.10) |
| **PGP** | 42.44 | 5.77 (0.02) | 19.50 (0.24) | 34.08 (0.03) | 34.40 (0.05) | 42.95 (0.12) |
| **BRAIN** | 119.66 | 8.91 (0.01) | 54.89 (0.26) | 113.41 (0.02) | 114.09 (0.06) | 122.27 (0.20) |

**Table 3.2** *Spectral gap between the largest and the second-largest eigenvalues of the adjacency matrix. For the dk-graphs, the shown values are the averages across different realizations for each d, while their standard deviations are reported in parentheses.*

| | Original | 0*k* | 1*k* | 2*k* | 2.1*k* | 2.5*k* |
|---|---|---|---|---|---|---|
| **INTERNET** | 17.56 | 0.70 (0.05) | 14.94 (0.53) | 18.83 (0.07) | 18.55 (0.11) | 19.53 (0.25) |
| **PGP** | 4.25 | 0.98 (0.04) | 5.51 (0.31) | 18.01 (0.18) | 17.55 (0.21) | 4.71 (0.19) |
| **BRAIN** | 40.97 | 2.90 (0.06) | 35.52 (0.31) | 77.53 (0.11) | 76.59 (0.27) | 42.71 (0.35) |

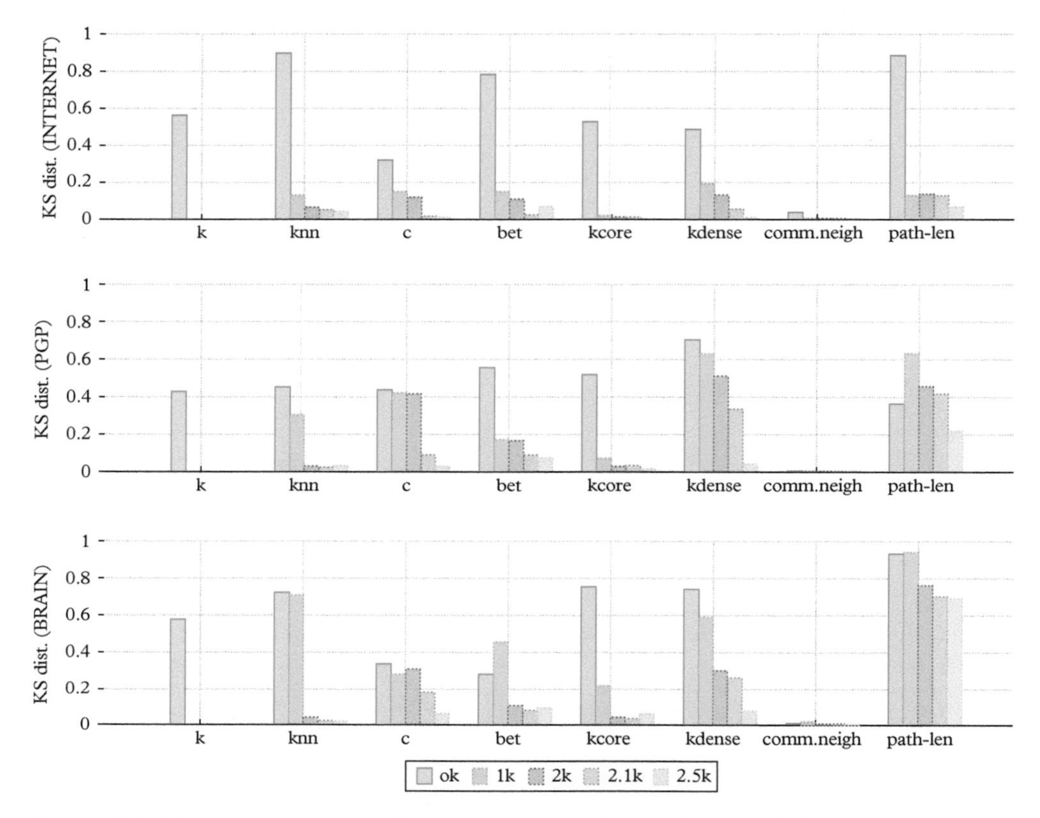


**Figure 3.7** *Kolmogorov–Smirnov distance between real networks and their* dk-*random graphs. The Kolmogorov–Smirnov (KS) distances between the distributions of per-node values of a given property in the real networks, and the same distributions in their dk-random graphs for the following properties: degree (k), average degree of nearest neighbors (knn), clustering coefficient (c), number of common neighbors (comm.neigh),* k-*coreness (kcore),* k-*density (kdense), betweenness (bet), and shortest-path distance (path-len). Adapted from Ref. [219].*

very close to zero for higher values of $d$, indicating that these two networks can be well approximated with $2k$- and $2.5k$-random graphs. On the other hand, for BRAIN, the global properties exhibit slow or no convergence at all, so it is an outlier, that is, its properties can be captured with $dk$-random graphs with $d \geq 3$. Although many properties can be reproduced with $2.5k$-graphs, we find that community structure is not preserved for any of these networks, regardless of the community detection algorithm.

## 3.3 *dk*-Series in multilayer networks

In this section, we discuss how *dk*-series generalize to multilayer networks in general and to multiplex networks in particular. The key idea behind the generalized *dk*-series is the same as in the case of monolayer networks: a *dk*-series is a series of inclusive and convergent statistics based on the frequencies of degree-labeled subgraphs of increasing size in a given network.

### 3.3.1 Multilayer networks

Here, we consider the most general case of multilayer networks with the most detailed form of generalized *dk*-statistics.

The structure of a multilayer network is fully specified by the adjacency tensor $A_{\alpha i,\beta j}$, where indices $\alpha, \beta = 1, \ldots, L$ indicate layers, while $i, j = 1, \ldots, N$ indicate nodes: $A_{\alpha i,\beta j} = 1$ if node $i$ at layer $\alpha$ is connected to node $j$ at layer $\beta$, and $A_{\alpha i,\beta j} = 0$ otherwise [72] (see also Chapter 1). To simplify the notation, from now on, we assume that networks are undirected and have no loops—$A_{\alpha i,\beta j} = A_{\beta j,\alpha i}$, and $A_{\alpha i,\alpha i} = 0$, for any combination of $\alpha$, $\beta$, $i$, and $j$—but the generalization to directed networks is straightforward.

The $0k$-statistics, that is, the number of edges in the network, and the $0k$-"distribution," that is, the average degree, are no longer scalar $M$ and $\bar{k}$ as in the monolayer case, but the $L \times L$-matrices

$$\hat{M} \equiv M_{\alpha\beta} = \sum_{i \leq j} A_{\alpha i,\beta j}, \tag{3.9}$$

$$\bar{\hat{k}} \equiv \bar{k}_{\alpha\beta} = \frac{2M_{\alpha\beta}}{N}, \tag{3.10}$$

specifying the number of interlayer edges between layers $\alpha$ and $\beta$ if $\alpha \neq \beta$, or the number of intralayer edges at layer $\alpha$ if $\alpha = \beta$. Similarly, the degree of node $i$ is no longer scalar but the $L \times L$-matrix

$$\hat{k}_i \equiv (k_i)_{\alpha\beta} = \sum_j A_{\alpha i,\beta j} \tag{3.11}$$

specifying the number of node's connections to other nodes in the same layer if $\alpha = \beta$, or to nodes in other layers if $\alpha \neq \beta$. If node $i$ is not present at layer $\alpha$, then $(k_i)_{\alpha\beta} = 0$. Node degrees are thus also matrices $\hat{k} = k_{\alpha\beta}$. A node that has this degree has $k_{\alpha\beta}$ connections from layer $\alpha$ to other nodes at layer $\beta$, which can be equal to $\alpha$. We note that the degree matrices are not, in general, symmetric, even if the network is undirected.

The $1k$-statistics, that is, the number of nodes $N(\hat{k})$ with degree $\hat{k}$, and the $1k$-distribution are then

$$N(\hat{k}) = \sum_i \delta(\hat{k}_i, \hat{k}), \tag{3.12}$$

$$P(\hat{k}) = \frac{N(\hat{k})}{N}, \tag{3.13}$$

where $\delta$ stands for the Kronecker delta, and the distribution is properly normalized: $\sum_{\hat{k}} P(\hat{k}) = 1$. In contrast with the monolayer case, the degree distribution is no longer a univariate distribution but a multivariate joint distribution of $L^2$ variables. In particular, this distribution contains all the information on the correlation of degrees of the same node at different layers. As in the monolayer case, the $1k$-distribution fully defines the $0k$-distribution via

$$\bar{\hat{k}} = \sum_{\hat{k}} \hat{k} P(\hat{k}). \tag{3.14}$$

The $2k$-statistics, that is, the number of links $N_{\alpha\alpha'}(\hat{k}, \hat{k}')$ between nodes of degrees $\hat{k}$ and $\hat{k}'$ at layers $\alpha$ and $\alpha'$, and the corresponding $2k$-distribution are given by the matrices

$$\hat{N}(\hat{k}, \hat{k}') \equiv N_{\alpha\alpha'}(\hat{k}, \hat{k}') = \sum_{i \leq i'} A_{\alpha i, \alpha' i'} \delta(\hat{k}_i, \hat{k}) \delta(\hat{k}_{i'}, \hat{k}'), \tag{3.15}$$

$$\hat{P}(\hat{k}, \hat{k}') = \mu(\hat{k}, \hat{k}') \frac{\hat{N}(\hat{k}, \hat{k}')}{2\hat{M}}, \text{ where} \tag{3.16}$$

$$\mu(\hat{k}, \hat{k}') = \begin{cases} 2, & \text{if } \hat{k} = \hat{k}', \\ 1, & \text{otherwise}, \end{cases} \tag{3.17}$$

is the factor taking care of proper normalization $\sum_{\hat{k}, \hat{k}'} P_{\alpha\alpha'}(\hat{k}, \hat{k}') = 1$ for any $\alpha, \alpha'$. Here and below, all vector, matrix, and tensor multiplication and divisions are element-wise, for example, $\left(\hat{N}/\hat{M}\right)_{\alpha\alpha'} = \hat{N}_{\alpha\alpha'}/\hat{M}_{\alpha\alpha'}$. Instead of a joint distribution of two variables $P(k, k')$ in the monolayer case, we deal with $L(L-1)$ joint distributions of $2L^2$ variables $P_{\alpha\alpha'}(\hat{k}, \hat{k}')$. These distributions contain strictly more information about degree correlations than the $1k$-distribution does. In particular, in addition to capturing the degree correlations of the same node ($i = j$) across layers, they also encompass all the degree correlations of distinct connected nodes ($i \neq j$) across both intralayer ($\alpha = \alpha'$) and interlayer ($\alpha \neq \alpha'$) connections. The $2k$-distributions define the $1k$-distribution similarly to the monolayer case:

$$P(\hat{k}) = \frac{\bar{\hat{k}}}{\hat{k}} \sum_{\hat{k}'} \hat{P}(\hat{k}, \hat{k}'). \tag{3.18}$$

It is evident from the expressions above that the *dk*-series in multilayer networks are different from the *dk*-series in monolayer networks only in that the scalar number of edges and node degrees are replaced by the $L \times L$-matrices $M \to \hat{M}$ and $k \to \hat{k}$, while their *dk*-distributions form tensors of rank $d$ whose indices are layers. If $d = 1$, this tensor is trivial: $\mathbf{P}(\hat{k}) \equiv P_\alpha(\hat{k}) = P(\hat{k})$ for any layer $\alpha$; but, starting with $d = 2$, any two elements of these *dk*-distribution tensors can, in general, be different, specifying in the $d = 2$ case, for instance, the degree correlations across pairs of different layers if $\alpha \neq \alpha'$, or within the same layer if $\alpha = \alpha'$. Each element of these *dk*-tensors is a joint distribution of $d$ degrees, that is, of $dL^2$ variables.

All higher-order statistics and distributions are then defined exactly as in the monolayer case, albeit with these two modifications. For instance, the $3k$-distribution is defined by Eqs (3.4)–(3.5) and determines the $2k$-distributions via Eq. (3.8), except that scalar degrees in these equations are replaced by degree matrices, and the numbers of wedges and triangles have three indices specifying to which layers the three nodes forming these two subgraphs belong. As in the monolayer case, higher-$d$ *dk*-distributions determine the degree correlations of nodes at distance $d-1$, the frequencies of $d$-cliques, including clustering at $d = 3$, and so on.

### 3.3.2 Multiplex networks

In node-aligned multiplex networks, all nodes are present in all layers, and all interlayer connections are trivial: every node is connected only to all its copies in all other layers, so that interlayer connections form $N$ disjoint $L$-cliques: $A_{\alpha i,\beta i} = 1$ for any combination of $i$ and $\alpha \neq \beta$. The *dk*-series can therefore be excused from keeping track of statistics of interlayer connections, which somewhat simplifies the formalism in the previous section, as described below. This simplification boils down to per-layer projections of the most detailed *dk*-statistics discussed in the previous section.

Since in multiplex networks $k_{\alpha\beta} = L-1$ for all nodes and all $\beta \neq \alpha$, all the off-diagonal components of degree matrices $\hat{k}$ are not informative and can thus be dropped, mapping degree $L \times L$-matrices to degree $L$-vectors composed of the diagonal elements of $\hat{k}$: $\hat{k} \mapsto \boldsymbol{k} \equiv diag(\hat{k})$. The $\alpha$-component $k_\alpha$ of this vector $\boldsymbol{k}$ specifies the number of intralayer connections of a node at layer $\alpha$. Similarly, the number-of-edges matrix $\hat{M}$ maps to the vector $\mathbf{M} \equiv diag(M)$, whose components $M_\alpha$ are the numbers of intralayer edges within layer $\alpha$.

Similarly, it is convenient to project the *dk*-distribution tensors per layer, forming vectors of distributions $\mathbf{P}(\boldsymbol{k},\boldsymbol{k}',\ldots) \equiv P_\alpha(\boldsymbol{k},\boldsymbol{k}',\ldots)$, consisting of the diagonal elements of the full distribution tensor $P_{\alpha,\alpha',\ldots}(\boldsymbol{k},\boldsymbol{k}',\ldots)$, that is, $P_\alpha(\boldsymbol{k},\boldsymbol{k}',\ldots) = P_{\alpha,\alpha,\ldots}(\boldsymbol{k},\boldsymbol{k}',\ldots)$, thus keeping track only of intralayer correlations of degrees of different nodes. The correlations of the degrees of the same node at different layers are still contained in $P(\boldsymbol{k})$.

Given this simplified representation, the *dk*-statistics, distributions, and their relations are exactly as in the general multilayer case, except that all matrices and tensors are replaced by vectors whose components are layers. For $d = 0, 1, 2$, for instance, we have the following expressions:

$$\bar{\boldsymbol{k}} = \frac{2\mathbf{M}}{N}, \tag{3.19}$$

$$N(\boldsymbol{k}) = \sum_i \delta(\boldsymbol{k}_i, \boldsymbol{k}), \tag{3.20}$$

$$P(\boldsymbol{k}) = \frac{N(\boldsymbol{k})}{N}, \tag{3.21}$$

$$\bar{\boldsymbol{k}} = \sum_{\boldsymbol{k}} \boldsymbol{k} P(\boldsymbol{k}), \tag{3.22}$$

$$\mathbf{N}(\boldsymbol{k}, \boldsymbol{k}') \equiv N_\alpha(\boldsymbol{k}, \boldsymbol{k}') = \sum_{i \leq i'} A_{\alpha i, \alpha i'} \delta(\boldsymbol{k}_i, \boldsymbol{k}) \delta(\boldsymbol{k}_{i'}, \boldsymbol{k}'), \tag{3.23}$$

$$\mathbf{P}(\boldsymbol{k}, \boldsymbol{k}') = \mu(\boldsymbol{k}, \boldsymbol{k}') \frac{\mathbf{N}(\boldsymbol{k}, \boldsymbol{k}')}{2\mathbf{M}}, \tag{3.24}$$

$$P(\boldsymbol{k}) = \frac{\bar{\boldsymbol{k}}}{\boldsymbol{k}} \sum_{\boldsymbol{k}'} \mathbf{P}(\boldsymbol{k}, \boldsymbol{k}'), \tag{3.25}$$

which all are lists of $L$ per-layer standard monolayer expressions, except that degrees are vectors. Compared to general multilayer networks, the *dk*-distributions in multiplex networks with these simplifications are all $L$-vectors, for $d > 1$, whose components consists of joint distributions of $d$ degrees, that is, of $dL$ variables.

### 3.3.3 Application to real networks

The general methodology behind the application of *dk*-series generalized to multilayer networks is the same as in the monolayer case discussed in Section 3.2. Yet, one has to keep in mind that multilayer *dk*-statistics tend to be extremely sparse and thus extremely constraining, even in the multiplex-projected case. This is because, compared to monolayer networks, the *dk*-distributions in multilayer networks contain much more detailed information about degrees, which are no longer scalars but matrices or vectors, and about their correlations within subgraphs of different sizes. Therefore, it is usually convenient to consider summary statistics of these distributions and define graph randomization procedures based on those. These procedures may depend on a particular choice of real network, on its specifics, and on particular questions one is to answer about the network.

For instance, in Ref. [79], generalized *dk*-series were applied to the Internet at the AS level. This network is a multiplex network with two layers. One layer consists of directed customer-provider links, for which, in order to send traffic over them, customer ASs must pay provider ASs, while the other layer consists of undirected peer-to-peer links connecting mostly large Internet service provider ASs, which exchange traffic free of charge over these links, based on bilateral agreements. The specific question addressed for this network was how to generate synthetic random graphs of varying sizes that reproduce specific types of degree correlations that reflect realities of business relationships between ASs in the Internet. For instance, peer-to-peer links tend to exist only between large Internet providers of large AS degree. Large providers tend to have large number of customers, a handful of peers, and few or no providers. Small customer ASs have no customers, no peers, and a small number of providers, and so on.

To properly capture these correlations, three joint distributions were considered. One was the full $1k$-distribution $P(\boldsymbol{k})$ specifying the correlations among the numbers of customer, provider, and peer connections that nodes have. Since the customer–provider layer is a directed network, this distribution is a joint distribution of three variables: in- and out-degrees $k_{1,in}$ and $k_{1,out}$ in the customer–provider layer, and degrees $k_2$ in the peer-to-peer layer. The other two distributions were the 2$k$-distributions of the total degrees $k = \sum_\alpha k_\alpha = k_{1,in} + k_{1,out} + k_2 = |\boldsymbol{k}|_1$ in the two layers. These distributions are projections of the multiplex 2$k$-distributions: $\mathbf{P}(k,k') = \sum_{\boldsymbol{k},\boldsymbol{k}'} \mathbf{P}(\boldsymbol{k},\boldsymbol{k}')\delta(|\boldsymbol{k}|_1,k)\delta(|\boldsymbol{k}'|_1,k')$. After these distributions were determined from data from the real Internet data, the *dk*-series were used to generate synthetic graphs of any size reproducing all the degree correlations contained in these distributions by first computing the marginals of these distributions $P(k_{1,in})$, $P(k_{1,out})$, $P(k_2)$, and $P(k)$, and the three copulas [206] representing their correlations in their joint distributions $P(\boldsymbol{k}) = P(k_{1,in}, k_{1,out}, k_2)$, $P_1(k,k')$, and $P_2(k,k')$. Joint degree sequences of varying lengths were then sampled from these copulas, and random graphs were constructed using stub-matching procedures. As expected, the degree correlations in these random graphs reproduced the degree correlations in the real Internet. Many other important structural properties of the Internet, including properties specific to the Internet, were reproduced by these 2$k$-random graphs as well [79], corroborating the finding that the Internet is nearly 2$k$-random with respect to many important properties [182, 219].

We conclude this section by reiterating that, when classifying, that is, determining "how random" a given multilayer or multiplex network is, the full matrix-degree-based *dk*-series provides a rich set of inclusive and convergent statistics, which contain a variety of summary statistics as different projections of the full *dk*-distributions. Any combination of these statistics constrained to their values observed in a given network defines a null random graph model, in which any structural property of the network can be tested on its typicality in the model. There seems to be no good-for-all-networks rule of what these projections are, as different multilayer networks may require different projections. Yet, the full joint 1$k$-distribution should most likely be always considered, while per-layer projections of *dk*-distributions with $d > 1$ are likely to be good projection choices for many networks, especially multiplex ones.

## 3.4 Discussion and conclusion

Topological measures commonly used for characterizing the structure of complex networks are interdependent, but the relation and the extent of their mutual correlations are often unknown. For this reason, systematic classification of networks via standard topological measures is not feasible. Here, we show how the problem of interdependence can be overcome by finding the set of base properties that can be used to explain all other relevant topological features of a network's structure. We describe the methodology proposed in Ref. [219] and show that most topological properties, which are deemed relevant for dynamics and function of networks, can be reproduced by random graphs with fixed degree distribution, degree-degree correlations, an average clustering coefficient and a degree-dependent average clustering coefficient, as in a given real network.

There is no reason to expect that non-local properties, namely, mesoscopic and macroscopic properties, cannot be reproduced by random graphs with local constraints.

And, for some networks and some properties, this is true. Our numerical experiments show that global features of brain networks, for example, the shortest-path length and betweenness distributions, differ drastically between the original network and *dk*-random graphs. This suggests that brain network evolution was subjected to some global constraints, which is reflected in its structure. The human brain consists of two weakly connected parts, corresponding to two brain hemispheres, a feature that cannot be reproduced with *dk*-random graphs with small $d$. In general, *dk*-random graphs with fixed local properties can not reproduce the community structure of all complex networks studied in Ref. [219], that is, the cluster organization is not robust to *dk*-randomization.

On the other hand, INTERNET and PGP, and other networks considered in Ref. [219], are clearly *dk*-random, with $d \leq 2.5$. Our analysis shows that the most basic properties of these networks, including microscopic, mesoscopic, and macroscopic ones, are a consequence of several local *dk*-properties: degree distribution, degree–degree correlations, and global and degree-dependent average clustering coefficients. This implies that the evolution of these networks was dominated by local dynamical rules and that it can be explained to a certain extent by mechanisms that are responsible for the manifestation of specific *dk*-properties. There already exists a multitude of approaches [80, 151, 302, 10, 225, 34] proposing different mechanisms to explain the emergence of these local topological properties. Clearly, the features that cannot be reproduced by *dk*-random graphs require separate explanations, or maybe some other set of base properties and different systems of null models.

The most basic topological features considered in this work can be considered non-significant, that is, there exists a *dk*-property captured by the corresponding *dk*-random graphs. In general, to tell how statistically significant a particular feature is, one needs to compare this feature in a real network with the same feature in an ensemble of random graphs, that is, a null model. The choice of the null model is free, but one should be careful when choosing the null model, since the significance of a certain feature is strongly dependent on it. The *dk*-random graphs discussed in this chapter can be used for determining the right network topology generator. One should first check whether most topological features of networks can be reproduced in *dk*-random graphs with a low value of $d$. If this is the case, then one may not need any sophisticated mission-specific topology generators. The proposed extension of *dk*-series to multilayer graphs enables the use of similar procedures on wider classes of real networks.

There are certain drawbacks of our approach that have to be mentioned. First, we do not have a proof that the proposed *dk*-random graph generation algorithms for $d = 2.1$ and $d = 2.5$ sample graphs uniformly at random from the ensemble. Second, it is known that the random graph ensembles and edge-rewiring processes employed here suffer from problems such as degeneracy and hysteresis [97, 239, 133]. The ideal solution would be to calculate analytically the expected value of given property in an ensemble. For this, we need an analytical description of null models which is currently only available for soft $d = 0, 1, 2$-random graph models [281, 282, 59, 58]. Unfortunately, the null models for generation of random graphs ensembles with constraints $d > 2$ are still not feasible, and they appear to be beyond the reach in the near future. We also lack algorithms for the generation of *dk*-random graphs for multiplex networks which would allow us to apply similar procedures and quantify the randomness of multilayer networks. Clearly, the solution of these problems will be of great importance for a full understanding of the relationship between the structure, function, and dynamics of real networks.

# 4

# Economic Specialization and the Nested Bipartite Network of City–Firm Relations

**Antonios Garas[1], Céline Rozenblat[2], and Frank Schweitzer[1]**

[1]Chair of Systems Design, ETH Zurich, Weinbergstrasse 58, 8092 Zurich, Switzerland
[2]Geography Institute, Faculty of Geosciences, University of Lausanne, 1015 Lausanne, Switzerland

The set of goods and services produced in a city depend on a complex interplay of factors that include institutions, taxes, skilled personnel, industrial heritage, and the presence of particular resources. Dependent on the availability of such factors, some cities have specialized in certain economic activities, while others became economically more diversified. Specialization comes with a benefit as it allows for economic multiplier effects through "agglomeration economies" [215, 132, 160]. This, however, can turn into a drawback if a particular economic activity goes into recession. Then, cities specialized in this activity will be distressed more than the economically diversified ones. To exacerbate the problem, in a globalized world the economic performance of a city increasingly depends on the economic performance of other cities. Such economic dependencies emerge even between cities that are very far away in terms of geographical distance, which, due to global economic linkages, are, in reality, proximal in economic terms [256].

This leaves us with the problem of quantifying such dependencies and linking them to the diversification of economic activities in cities. In this chapter, we identify a city's economic activities by monitoring firms with global presence and which operate in a particular city. This way, we can create a network that links cities and firms and, if we focus on the economic activities of each firm, we can extend this network by linking cities to economic activities. Networks that describe realtions between different sets of nodes, in our case, cities and economic activities, are called bipartite networks (or bipartite graphs) [78]. The bipartite network we construct to link every city with the economic

Garas, A., Rozenblat, C., and Schweitzer, F., "Economic Specialization and the Nested Bipartite Network of City–Firm Relations" in *Multiplex and Multilevel Networks*, edited by Battiston, S., Caldarelli, G., and Garas, A. © Oxford University Press 2019.
DOI: 10.1093/oso/9780198809456.003.0004

activities of firms with global presence will allow us to study *how specialized or diversified* each city is with respect to a global context.

In general, firms are connected to other firms through different types of links, such as ownership relations, supply chains, and financial obligations. Likewise, cities are linked to other cities via transportation networks or financial networks. Therefore, we can represent all these relations using large, multilayered, interconnected network [107], where the different layers contain information about different kind of links connecting cities to cities, firms to firms, and cities to firms. Indeed, this is the most general way to describe any system with different interacting elements, using the complex networks framework. In this representation, the aforementioned bipartite network is just the part that describes the structure of the interlayer links that connect cities to firms, based on the firm's locations.

The structure of the city–firm bipartite network has striking similarities with other types of bipartite networks found in *ecology*. There, nodes represent species, while links their interactions. In so-called antagonistic networks, the interaction between species is asymmetric, such as host–parasite, predator–prey, and plant–herbivore interactions. In so-called *mutualistic* networks, on the other hand, the interaction between species is symmetric, that is, both species interact in a mutually *beneficial* way, such as, for example, the way that plants interact with their pollinators. Networks with antagonistic and mutualistic interactions have long been studied in ecology, to show that the stability of ecological communities is linked to structural features of the network topology [24, 293, 240]. More precisely, it was shown that mutualistic networks are organized in a nested pattern, while antagonistic networks are organized in compartments [293]. A nested organization means that the network consists of sets of generalist nodes and sets of specialist nodes. The specialists interact only with a small subset of nodes, while the generalists interact with (almost) all other nodes in the network. In nested ecosystems, the large set of interactions between generalists (i.e., species that interact with many other species) creates a dense core to which the specialists (i.e., species that interact with few other species) are attached. It was shown that the (empirically observed [21]) nested structure of mutualistic networks reduces the interspecies competition, which, as a consequence, allows ecosystems to support more species and increase biodiversity [24].

Thus, it would be of great interest to see whether such nested structures can be also found in bipartite networks related to economics. A recent study [129, 288] has investigated the bipartite network between firms and countries, to relate it to economic stability. It was found that robust countries have, indeed, a wide range of diversification in their economic activities. Based on this economic complexity, performance measures for countries were proposed. Specifically, it was shown that the dynamics of the nested structure of industrial ecosystems can predict path dependencies in the way industries appear and disappear in given countries. This helped to explain the evolution of the set of products that are produced and exported by these countries [248, 52]. A different analysis, focused on the New York garment industry, showed that a firm's survival probability depends on the firm's position in the nested network of interactions between designer and contractor firms [248].

In this chapter, however, we are interested in the bipartite network linking cities and the economic activities of globalized firms. These interactions are mutualistic because cities benefit from the operation of firms through taxes, employment, and so on, while firms benefit from cities through access to infrastructure, resources, customer bases, skilled personnel, and so on. To build our bipartite network of city–economic activity relations (an extract of which is shown in Figure 4.1), we used data about firms with global presence. For these firms, we know the precise locations of the headquarters and all their subsidiaries and, using the standard *Nomenclature of Economic Activities* (NACE), we can assign their core business to an economic sector (for details, see Section 4.1). Next, we studied the *structure* of this network, and we show that it follows a nested assembly, similar to ecological mutualistic networks. Therefore, building upon previous works in the field of ecology and their follow-ups with respect to economic networks, we show that ecological indicators can be used to identify the unbalanced deployment of economic activities, and we provide evidence that the structure of this bipartite network of city–firm relations contains information about the quality of life in cities.

## 4.1 The nested structure of links between cities and economic activities

For our analysis, we used data about the 3,000 largest firms with global presence and their $\sim$1 million direct and indirect links to $\sim$800,000 subsidiaries extracted from the BvD *orbis* database of 2010 [51]. The firm locations were aggregated using the concept of Functional Urban Areas (FUAs), which was developed by the *European Spatial Planning Organization Network* (ESPON) [87]. Using FUAs makes it possible to agglomerate municipalities according to their functional orientation—sometimes going beyond administrative boundaries—and reflect the actual operational conditions of people, enterprises, and communities. Therefore, FUA agglomerations result in an efficient mapping of the economic activity and service production. In addition, we classified firms into economic sectors according to their core businesses, using the NACE nomenclature provided by *eurostat* [89]. These sectors were further aggregated following the United Nation's *International Standard Industrial Classification of All Economic Activities* (ISIC REV.4) methodology, which results to an aggregation of activities into 21 different sections. Pairing the geographic location given by the FUAs with the NACE-ISIC classification of every individual firm, we created a bipartite network of interactions between 1,169 cities and 21 economic activities. An example of this network for the ten cities with the largest number of firms is shown in Figure 4.1.

Our bipartite network is represented by an incidence matrix $M$ pairing each of the 21 economic activities to each of the 1,169 cities where this activity is present (see Figure 4.2). Therefore, each matrix element $m_ij$ has a value of 1 if the economic activity $i$ is present in the city $j$, and 0 otherwise.

To calculate the nestedness value of this matrix, we used the NODF algorithm developed by Almeida-Neto *et al.* [7]. This algorithm returns a nestedness value $N$ in the range [0, 100], with $N = 0$ when there is no nestedness, and $N = 100$ for the case of perfect nestedness. To assess the significance of nestedness, we compared our measured value with a benchmark null model. In this chapter, our model of choice is the null model introduced by Bascompte *et al.* [21], which creates randomized networks by preserving

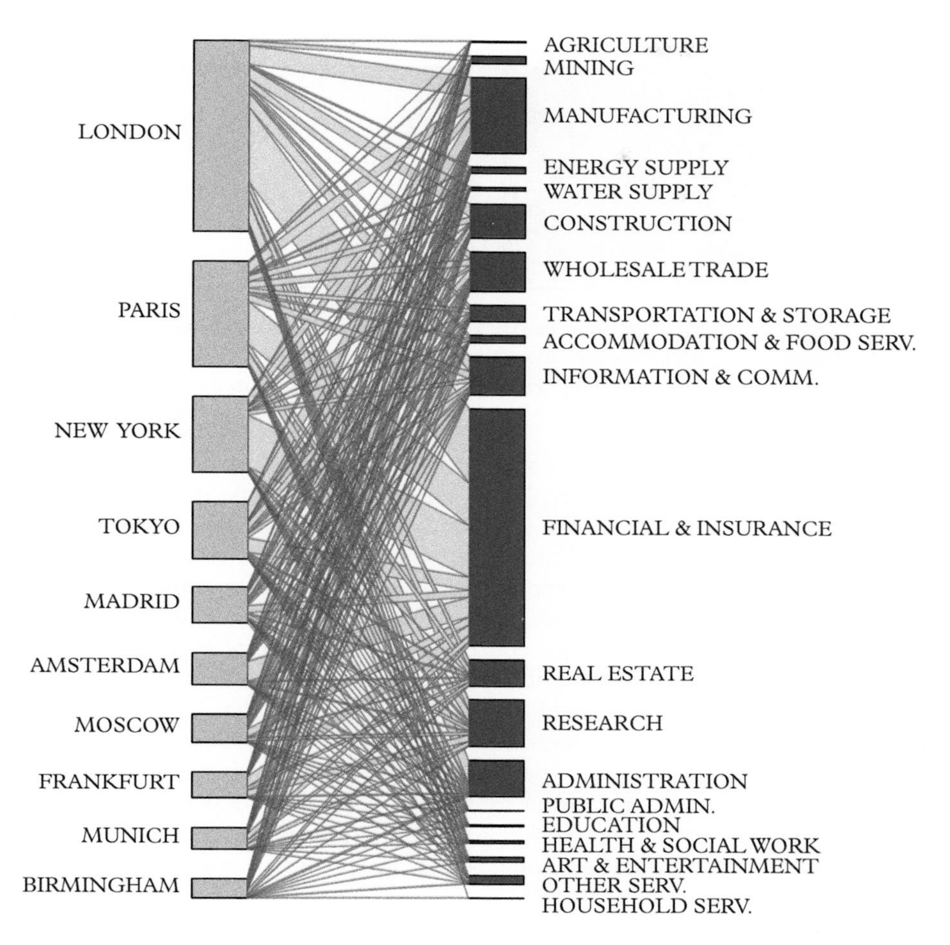


**Figure 4.1** *City–economic activity mutualistic interactions. City–activity interactions as a bipartite network. The activity links of the ten cities with the largest number of firms are shown. The link width between city* i *and activity* j *corresponds to the number of firms associated to activity* j *located in city* i.

the degree distribution of the original network. This model generates ensembles of swapped matrices $\tilde{M}$ where the probability of each matrix cell being occupied is the average of the probabilities of occupancy of its row and column. Practically, this means that the probability of drawing an interaction is proportional to the level of generalization (degree) of both the city and the economic activity, that is $p_ij = (k_i/n_j + k_j/n_i)/2$, where $k_j$ is the degree of the city, and $k_i$ is the degree of the activity in the bipartite network, while $n_i$ and $n_j$ are the number of available activities and cities, respectively.

To measure the contribution of each individual city to the nestedness value of the whole network, we follow the methodology of Saavedra *et al.* [248]. More precisely, we calculate $c_i = (N - \langle N_i^* \rangle)/\sigma_{N_i^*}$, where $N$ is the observed nestedness of the whole network,

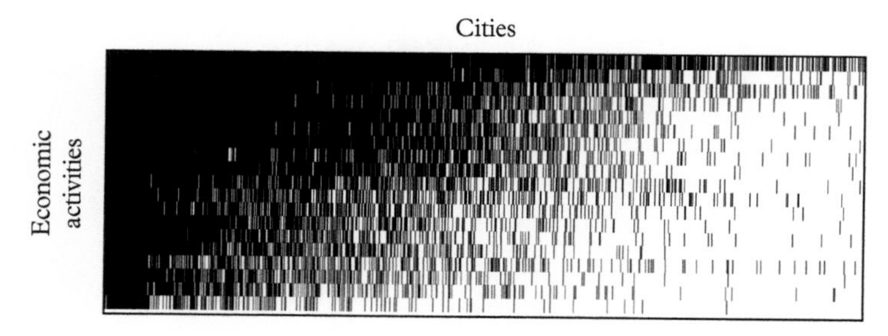


**Figure 4.2** *The city–activity interaction matrix Plot of the interaction (incidence) matrix* M *pairing each of the 21 economic activities to each of the 1,169 cities where this activity is present. Each matrix element* $m_{ij}$ *has the value 1 if the economic activity* i *is present in the city* j*, and 0 otherwise.*

and $\langle N_i^* \rangle$ and $\sigma_{N_i^*}$ are the average and standard deviation, respectively, of the nestedness across an ensemble of 100 random replicates for which all the links of city $i$ to economic activities have been randomized. The number of random replicates is chosen in order to provide optimal performance, while at the same time the individual contribution to the nestedness of each city has converged significantly to their asymptotic value. More precisely, we performed a convergence analysis for which we calculated the Spearman's $\rho$ correlation coefficient between two consecutive rankings of cities according to their nestedness contribution, with increasing numbers of replicates. From this analysis, we (a) observed that the rankings indeed converge to a saturation level and (b) concluded that 100 random replicates are enough, as $\rho$ is already almost 0.99.

Using this methodology, we find that the bipartite network of cities–economic activities is nested (see Figure 4.2), with a nestedness value $N = 78.4$ ($p < 0.0001$). This already highlights structural similarities in the interaction patterns that occur in a natural ecological system and in the human-made economic system. And since a nested network structure is known to promote community stability in mutualistic ecological networks [293], we anticipate that the mutualistic network of cities and economic activities would be stable as well.

However, as was shown recently for both ecological and socioeconomic networks, nestedness comes at a price [248]. On the one hand, the nodes that contribute more to the nestedness of the network are the nodes that contribute more to the network persistence. On the other hand, these same nodes were identified as the ones most vulnerable to going extinct. Of course, in our case, a city might not go extinct, but it may decline to a less prosperous state.

## 4.2 Learning from ecology: Contribution to nestedness and economic well-being

As shown in Figure 4.3(a), the distribution of the individual contribution to nestedness is concentrated around the mean value $\mu = 1.96 \pm 0.01$. Therefore, one important question

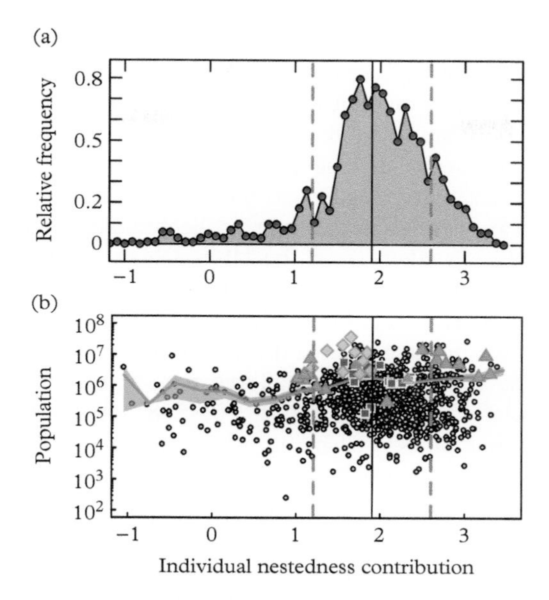


**Figure 4.3** *Individual nestedness contribution. (a) Distribution of individual nestedness contribution for all cities. (b). Population versus individual nestedness contribution. The red line shows the mean value, and the band shows the standard error. The blue squares highlight the location of the top ten cities according to the Mercer 2012* Quality of Living *worldwide city rankings [191] and the top ten cities according to the EIU's 2013* Global Liveability Ranking and Report *[292] while the green triangles highlight the location (where available) of the bottom ten cities for both these rankings. In addition, the orange diamonds indicate the locations of the ten cities with the largest numbers of firms.*

is whether this nestedness value has any relation to a city's economic performance. If it does, then where are the best performing cities located in this distribution? Are they close to the center or close to its tail? Unfortunately, we do not have access to data about economic performance for individual cities. However, we do expect economic performance to be strongly correlated with the well-being of a city's inhabitants. In this sense, using the rankings provided by the Economist Intelligence Unit's (EIU) 2013 *Global Liveability Ranking and Report* [292] and the Mercer 2012 *Quality of Living* worldwide city rankings [191], we calculated the nestedness contribution of the top ten and the bottom ten cities. We found that, in both ranking systems, all of the top cities are within the range of $\mu \pm \sigma$, while 70% of the bottom cities are outside this range. More precisely, 41% of them are above $\mu + \sigma$, and 39% below $\mu - \sigma$. From Mercer's bottom-ten list, the cities above $\mu + \sigma$ are Abidjan, Khartoum, Kinshasa, and Conakry, while, from EIU's list, Karachi, Algiers, and Douala are in this range. In the area below $\mu - \sigma$, from Mercer's bottom-ten list we find Tbilisi, Sanaa, and Baghdad and, from the EIU's list, Damascus and Tripoli. This discussion shows that a ranking based on the nestedness score gives insightful results, where the better performing cities, according to Mercer and the EIU [191, 292] are closer to the mean of the nestedness distribution, while the worst performing ones are further away (see Figure 4.3(b)).

Given the general tendency of cities to grow, it is natural to ask if there is any measurable impact of population to the nestedness score. It is already known that a city's population drives many diverse properties of cities [31]. Are smaller cities more stable or more vulnerable, according to the way stability/vulnerability is reflected through nestedness? To answer this question, we collected data about cities' populations, by consolidating information taken from the United Nations database on cities,[1] the Organisation for Economic Co-operation and Development's database on cities,[2] and the ESPON project [87]. As shown in Figure 4.3(b), there is no pronounced relationship between the (logarithm of) population and individual nestedness. The Pearson correlation coefficient $r = 0.069$ ($p = 0.017$) is small and not significant, and the same is true for the Spearman correlation $\rho = -0.0052$ ($p = 0.8588$). Of course, if a city performs well and increases its inhabitants' well-being, it may become the target of internal or external migration flows and eventually increase its population. However, its network position—as measured with respect to nestedness—does not seem to be influenced by the population.

We can, of course, anticipate that if a city is specialized in an economic activity, it will prosper as long as the activity fares well. If this activity is hit by turmoil, or just underperforms with respect to other activities, this may lead to the city's decline. To avoid this, diversification of activities is required; but how much diversification is enough? And even if a city has indeed diversified its activities, how does this diversification compare to that of other cities? It is expected that large cities are able to attract many firms that would populate multiple economic sectors of activity [234]. This means that large cities are by definition "generalists" in the bipartite graph, and this introduces a bias in our interpretation of the nestedness score. To be more specific, let us consider the case of Detroit, which has the nestedness score $c = 1.71$, which places it near the mean of the nestedness distribution. Based on this number alone, we would argue that Detroit performs well, and we would not have anticipated its bankruptcy on July 18, 2013. Therefore, it is not enough to ensure that multiple economic sectors are populated; it is also important to monitor how many firms populate each sector, which is equivalent to using information about the weights of the links in the bipartite network. If the distribution of firms in economic sectors is skewed, one or a few sectors will dominate. Thus, a major decline in the dominating sector will have a major impact on the city's economy, and this will indirectly affect all the other sectors as well.

In addition, coming back to our previous discussion on the multilayered interconnected structure of the overall city–firm network, this effect will be even more pronounced in the presence of links between firms from different sectors. Even though we do not have data about such "hidden" links, it is not hard to imagine that many service-related smaller firms (e.g., subcontractors or advance production services) provide

[1] http://data.un.org.
[2] http://stats.oecd.org/Index.aspx?DataSetCode=CITIES.

support and depend on the function of the large firms that belong to the dominant economic sector. Therefore, the decline of this sector will create a cascading effect that is very hard to properly evaluate in the absence of detailed dependency data.

In the example at hand, from the 4,455 total large firms that were active in Detroit, 2,299 belonged to the manufacturing sector. The second most populated sector was Financial & Insurance, with 642 firms. We expect that many of these firms have strong ties to the manufacturing companies and would be affected if something went wrong in the manufacturing sector. However, since we cannot document these ties, for simplicity, we will assume that all sectors are independent.

Hence, to detect when one sector is overrepresented in the overall economic activity, we calculated the fraction $f_\tau$, that is, the number of firms in the largest sector over the number of firms in all other sectors. If $f_\tau \leq 1$, the city is well diversified across activities, while if $f_\tau > 1$, a particular sector dominates the economy, and the city might be at risk. For our dataset, and under the assumption of sectoral independence, $f_\tau = 1.066$ for the case of Detroit, which indicates the city's vulnerability.

If more refined data were available, we could improve the calculation by dividing the number of firms of all sectors that would be significantly affected by a decline in the largest sector by the number of firms that would not be affected by this decline. However, such calculation cannot be performed with the datasets currently available. We would like to contrast our measure $f_\tau$ with other existing indexes for diversity. A well-known index used in ecology to quantify the biodiversity of a habitat is Simpson's index [266]. It is also known as the Herfindahl–Hirschman index in economics, where it is used to measure market concentration [130, 127]. We calculated Simpson's index for all cities in our database and found that the resulting concentration ranking is strongly correlated (Spearman's $\rho = -0.976$ ($p \leq 0.0001$)), with the ranking based on $f_\tau$. But most available indexes, including Simpson's index, do not allow for the easy identification of a threshold value that discriminates well-diversified cities from those that are not well diversified. This, however, can be achieved by the $f_\tau = 1$ value in our case.

There is a limitation when applying our $f_\tau$ index to extremely specialized cities, as it diverges in cases where (mostly due to data limitations) only one economic sector is present. These cases are also identified by Simpson's index as being extremely specialized cities and are assigned a zero value. It is, therefore, better practice to exclude such pathological cases from our analysis. For this reason, we restricted our calculation of $f_\tau$ to the 100 cities with the largest number of firms in our database.[3]

As shown in Figure 4.4, most of the cities have a $f_\tau < 1$, which is evidence of a balanced development. However, there are some cities with $f_\tau$ values not only $> 1$ but even greater than the value for Detroit. As it happens, most of these cities, which include New York, Amsterdam, Zurich, and so on, are large financial centers, and this fact highlights the fragility of an economic model that is largely dependent on

[3] We calculated the Simpson's index for the set of 100 cities with the largest number of firms in our database, and again we found that the resulting concentration ranking is strongly correlated (Spearman's $\rho = -0.963$ ($p \leq 0.0001$)) with the ranking based on $f_\tau$.

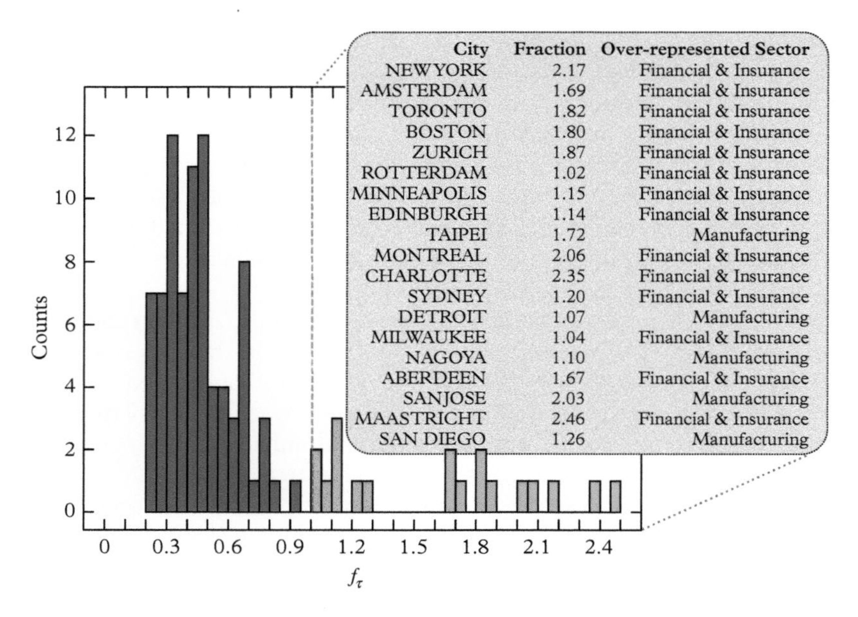


**Figure 4.4** *Concentration of economic activities. Histogram of the fraction $f_\tau$ for the 100 cities with the largest number of firms in our database.*

financial services. The recent financial crises rang some bells, and now policy makers in developed countries are trying to mitigate this dependency by reshoring manufacturing. A profound example of this is the EU's 10|100|20 strategy, which aims to get almost 20% of semiconductor manufacturing back to Europe by 2020 through an unprecedented public–private investment partnership.[4]

## 4.3 Conclusion

In summary, by using the multilayered network approach and exploiting similarities in the bipartite network describing the organization of interlayer links across complex systems, we can use indicators developed in ecology to assess the performance of a city in the globalized economy. With these indicators, we go beyond the mere evaluation of the economic specialization of cities [134], as we associate the specialization of a city to the vulnerability of the whole "ecosystem" describing city–economic activities relations, similar to the way the extinction of one species affects the stability of natural ecosystems. Therefore, we highlight the possibility that such indicators have the potential to identify the need for new multilevel policies able to regulate the cities at the national or continental

[4] http://www.semi.org/eu/node/8506.

level (such as within EU), in order to enhance their position in the bipartite network of city–economic activities relations.

However, there is also a need for closer supervision to prevent the overrepresentation of some economic activities at the expense of others, as this increases risk in the future. In this respect, policy interventions that reduce the dominance of one sector over others should be applied more frequently. Currently, the financial sector is strongly overrepresented in most large cities in developed countries; hence, policies like the EU 10|100|20 strategy are important to hedge against future risk.

## ACKNOWLEDGMENTS

The authors acknowledge financial support from the EU-FET project MULTIPLEX 317532. We thank Antoine Bellwald and Faraz Ahmed Zaidi for cleaning and aggregating the data. For our analysis, we used the R software for statistical analysis v3.0.2, and the bipartite library v2.02.

# 5

# Multiplex Modeling of Society

**János Kertész[1,2,3], János Török[1,2], Yohsuke Murase[4,5], Hang-Hyun Jo[6,7,3], and Kimmo Kaski[3]**

[1]CEU, Department of Network and Data Science, Nádor u. 9, Budapest, H-1051, Hungary
[2]BME, Institute of Physics, Budafoki út 8, H-1111, Hungary
[3]Department of Computer Science, Aalto University, P.O. Box 15500, Espoo, Finland
[4]RIKEN Advanced Institute for Computational Science, Kobe, Hyogo 650-0047, Japan
[5]CREST, Japan Science and Technology Agency, Kawaguchi, Saitama 332-0012, Japan
[6]Asia Pacific Center for Theoretical Physics, Pohang 37673, Republic of Korea
[7]Pohang University of Science and Technology, Pohang 37673, Republic of Korea

## 5.1 Introduction

Networks of social interactions are paradigmatic examples of multiplexity. It was recognized long ago by social scientists [94, 307] that the best way to interpret the network of different kinds of human relationships is a multiplex network, where each layer corresponds to a particular type of relationship, for example, between kin, friends, or co-workers (see [47] and references therein).

Until recently, only rather small networks could be studied, due to the limited size of datasets collected by traditional methods used in sociology [30]. Consequently, fundamental questions, such as what is the structure of the network of interactions at the societal level, could hardly be approached. In fact, the global consequences of local rules like those formulated in the famous Granovetter hypothesis [123] about the strength of weak ties could not be tested by the traditional methods. Over the past 15 years, this situation has changed substantially due to the large scale of human sociality–related datasets becoming increasingly available.

Social interaction between people can always be considered as a kind of communication. In the digital era, much of the communication has shifted to ICT channels, where records are created about all interactions. Mobile phone calls, text messages, social network services (SNSs) like Facebook and Twitter, and even massively multiplayer

Kertész, J., Török, J., Murase, Y., Jo, H.-H., and Kaski, K., "Multiplex Modeling of Society" in *Multiplex and Multilevel Networks*, edited by Battiston, S., Caldarelli, G., and Garas, A. © Oxford University Press 2019.
DOI: 10.1093/oso/9780198809456.003.0005

online games produce a deluge of data, which can be considered to constitute the digital footprints of individuals and thus serve as a gold mine for research into human sociality. Thanks to this development, a new discipline has emerged: computational social science [168].

Call detail records (CDRs) of mobile phones play a special role among datasets from today's communication tools [35], as the coverage is close to 100% in developed countries, and most people do not take a step without their devices. The CDRs completed with metadata like gender, age, zip code, and information about location open up further research possibilities. Such data were used, among others, to prove the Granovetter hypothesis [218], uncover regularities in human mobility patterns [121], and deduce the distance dependence of social ties [166]. Using the metadata, it was also possible to distinguish between different types of relations and relate the activities to the age and gender of the individuals [222, 141]. A large amount of observations have accumulated reflecting various interesting features of human interactions at the societal level [217, 35]. Many findings in the CDR dataset were found also to be characteristic of other communication channels, for example, email [156], Facebook [174], and Twitter [204]. Such features include broad distributions of network quantities like degree and weight (to be defined later), community structure, and assortative mixing. This way, a set of stylized facts have emerged [201], and they serve as guidelines for large-scale modeling of the society.

Society can be considered to be a multiplex, not only with respect to the different types of relationships but also from the point of view of the channels used for communication, such as face-to-face interaction, mobile phones, and SNSs. For the latter case, the layers of the multiplex correspond to the different communication channels. Figure 5.1 illustrates these two different ways of considering multiplexity.

A true picture about the entirety of human communication in the society should be based on comprehensive data from all the levels of this second type of multiplex. However, this is not feasible because, even in the digital era, not all forms of communication are registered. Moreover, data is usually available only for one channel, meaning that, in the whole multiplex, there is only one layer at our disposal for investigation. Linking data from diverse channels would, of course, be desirable but it is, in most cases, impossible, due to the different origins of the data and/or for privacy reasons.

Here, we will discuss aspects of multiplexity in modeling society. This chapter is organized as follows. First, we sum up the "stylized facts" as obtained from so-called Big Data. Then we show how the Granovetterian structure, identified in single-layer data, can be modeled in a multiplex setup and how this structure can coexist with overlapping communities as they naturally emerge. In the next section, we report on modeling channel selection to analyze the sampling bias introduced by single-channel data. Finally, we discuss the results and the outlook for the future.

### 5.1.1 Stylized facts for social networks

In recent years, the availability of a large number of digital datasets have enabled us to characterize the structure of social networks in more detail and up to an unprecedented

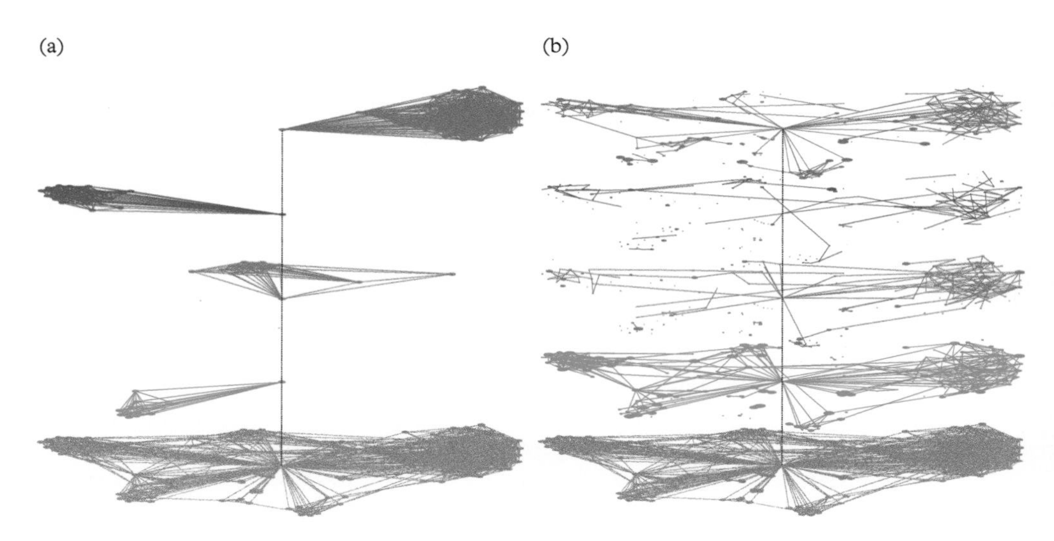


**Figure 5.1** *Egocentric network of a person (central vertical lines in the figure) as taken from the iWiW dataset, which, for simplicity, we assume to map out completely the person's relationships. This social network is resolved in two multiplex networks: (a) The layers correspond to different types of relationships or contexts, as obtained from combination of metadata and community detection. Only the four most important relationships are shown (four layers from the top). In (b), the layers represent communication channels; the top four are shown. In both cases, the bottom layer is the same and is the aggregate or projection of the multiplex, containing all the all contacts.*

scale. For example, researchers have investigated emails [83, 156], mobile phone calls [35, 218], datasets from SNSs like Facebook [298] and Twitter [165], and even data from face-to-face interactions [315, 135]. Analyses of such datasets revealed several commonly observed features or *stylized facts* for social networks, as summarized in Table 5.1. Here, we will mostly rely on the empirical findings from large-scale mobile phone call datasets [217] because, due to their large coverage, they are expected to reflect the features of real social networks to a large extent.

The most apparent stylized fact is the broadness of the distributions of the network quantities, such as the degree $k$, the link weight $w$, and the node strength $s$ [5, 217]. The weight of a link quantifies the interaction activity between two nodes. The strength, defined as the sum of weights of links involving the node, typically quantifies the activity of that node. The distributions of these quantities, that is, $P(k)$, $P(w)$, and $P(s)$, have been found to be not only broad but also overall decreasing, implying that individual and interaction activities are heterogeneous and the maximum of the distribution is at $k \approx 1$. The latter is clearly not consistent with the observation that, in a society, it is hard to find a person with only one or a few relationships. This discrepancy should be attributed to sampling effects, which will be discussed later in this chapter. The overall decreasing $P(w)$ can be interpreted as the prevalence of weak links or weak ties in social networks.

**Table 5.1** *Stylized facts in the CDR dataset compared to the expected behavior for the entire social network, adopted from [201]. The arrows indicate the general trend of the profile: ↗ (↘) implies that the profile is monotonically increasing (decreasing). The initially increasing and then decreasing behavior is denoted by ↗↘. The definitions of quantities are described in the main text.*

| | **CDR** | **Expected Behavior** |
|---|---|---|
| $P(k)$ | ↘ | ↗↘ |
| $P(s)$ | ↘ | ↗↘ |
| $P(w)$ | ↘ | ↘ |
| $s(k)$ | ↗ | ↗ |
| $k_{\mathrm{nn}}(k)$ | ↗ | ↗ |
| $O(w)$ | ↗↘ | ↗ |
| $c(k)$ | ↘ | ↘ |

Homophily is one of the main organizing principles of tie formation in social networks [190] as people tend to get along with those who have similar characteristics. Here, we are interested in the structure of social networks; thus, we focus on the degree–degree correlation. This correlation has been quantified in terms of assortativity, which can be measured by the Pearson correlation coefficient between degrees of neighboring nodes [208]. Many social networks are found to be assortative. A simple way to detect assortativity is to measure the average degree of neighbors for nodes with degree $k$, denoted by $k_{\mathrm{nn}}(k)$. An increasing trend means assortativity, as found, for example, in the CDR dataset [217].

High clustering is evident in social networks, as explained by the saying "friends of friends get friends." It means that if $B$ and $C$ are both connected to $A$, there is high chance that they are also connected to each other. The local clustering coefficient of a node is measured as the number of links between its neighbors, divided by the maximal possible number of such links. The average local clustering coefficient for nodes with degree $k$, denoted by $c(k)$, is found to be generally a decreasing function of $k$ (e.g., see [217]).

How individuals distribute their limited resources, like time, among their neighbors is also indicative when characterizing social networks. For this, the egocentric network, consisting of a node and its neighbors, has been studied in terms of the ranks of link activities or weights. A layered structure in the activity–rank relation has been reported [81], while smooth functional forms have been seen to fit with the empirical observations of single-channel data [279, 254].

Finally, on the mesoscopic scale of social networks, we find a rich community structure. This means that nodes in communities are densely connected, while nodes between different communities are sparsely connected [96]. This picture is important to account for large clustering in sparse social networks with inhomogeneous degree distribution where high-degree nodes or hubs occur. Such a topological property is correlated with link activity in that the communities of strongly connected nodes are weakly connected to each other [218], in agreement with the famous Granovetter's hypothesis [123]. Link-level consequences of weight–topology correlation can be measured by the average overlap for links with weight $w$, denoted by $O(w)$. The overlap of a link is the number of common neighbors of nodes connected by the link, divided by the total number of neighbors of those nodes. It has been found that the stronger links show larger overlap [218], up to 95% of the weights, thus showing agreement with Granovetter's hypothesis.

It should be emphasized that these stylized facts have been deduced from single-layer data, representing one layer of the multiplex in Figure 5.1(b). A single layer may reflect multiplex properties stemming from different types of relationships, as depicted in Figure 5.1(a), while this restriction introduces some bias, as we will show later in this chapter.

## 5.2 A weighted multilayer model

In order to reproduce the stylized facts shown in the previous section, a simple model was proposed by Kumpula *et al.* [162], which we will call the weighted social network (WSN) model. This model succeeded in reproducing various stylized facts, including community structure, the Granovetterian weight–topology relationship, assortative mixing, decreasing clustering spectrum, and the relationship between node strength and degree. However, the WSN model has only a single layer; thus, important aspects of the multilayer structure of social networks are missing.

As discussed in the introduction, people are involved in different social contexts or relationships, and their social networks are strongly context dependent [139, 140]. To handle this, social networks should be represented as multilayer networks or multiplexes [149, 37, 138], where links in the different layers correspond to different contexts (see Figure 5.1(a)). These contexts are hardly distinguishable from the available data; thus, observed networks should be considered to be aggregates of multiple layers. It is therefore challenging to construct a model that both reflects observations and has a multilayer structure. In the following, we discuss how to generalize the WSN model [202] and show what conditions are needed to reproduce both the Granovetterian weight–topology relationship and the community overlap arising from the multiplex nature of social networks.

Let us first summarize the original WSN model [162]. It considers an undirected weighted network of $N$ nodes. The links in the network are updated according to the following three rules. The first rule is called *local attachment* (LA). Node $i$ chooses one of its neighbors $j$ with probability proportional to $w_{ij}$, which stands for the weight of the link between nodes $i$ and $j$. Then, node $j$ chooses one of its neighbors except $i$, say $k$, randomly with probability proportional to $w_{jk}$. If nodes $i$ and $k$ are not connected, they get connected with probability $p_\Delta$ by a link of weight $w_0$; if they have already been connected, the weights of the links in the $(ijk)$ triangle, namely, $w_{ij}$, $w_{jk}$, and $w_{ik}$ are increased by $\delta$. The second rule is *global attachment* (GA), where a node is connected to a randomly chosen node with weight $w_0$. This happens with probability 1 if the node has no links; otherwise, it occurs with probability $p_r$. Finally, the third rule, *node deletion* (ND) is that, with probability $p_d$, a node can lose all its links. LA, GA, and ND are applied to all nodes at each time step, and we obtain a statistically stationary state after a sufficient number of updates. A snapshot of a network generated by this model is shown in Figure 5.2(a).

It is clear by visual inspection that the single-layer WSN model does not generate a significant number of overlapping communities. This is a consequence of the LA rule. Even if one node happens by chance to belong to two communities, such communities tend to be connected by the links created with LA, including the bridging node. While the LA mechanism is crucial for generating community structure, it tends to merge communities. Thus, a mechanism for keeping overlapping communities from being separated must be incorporated to reproduce overlapping communities found in reality. One simple and plausible way of modeling this is the introduction of the multilayer structure, as this is the main cause of the communities overlapping.

### 5.2.1 An uncorrelated multilayer WSN model

In order to study multilayer effects, we first generalize the single-layer WSN model in a naive way as follows. We consider $L$ layers of the same set of nodes and assume that each layer corresponds to a different type of relationship or communication context. For each layer, we independently construct a network in the same way as in the original single-layer

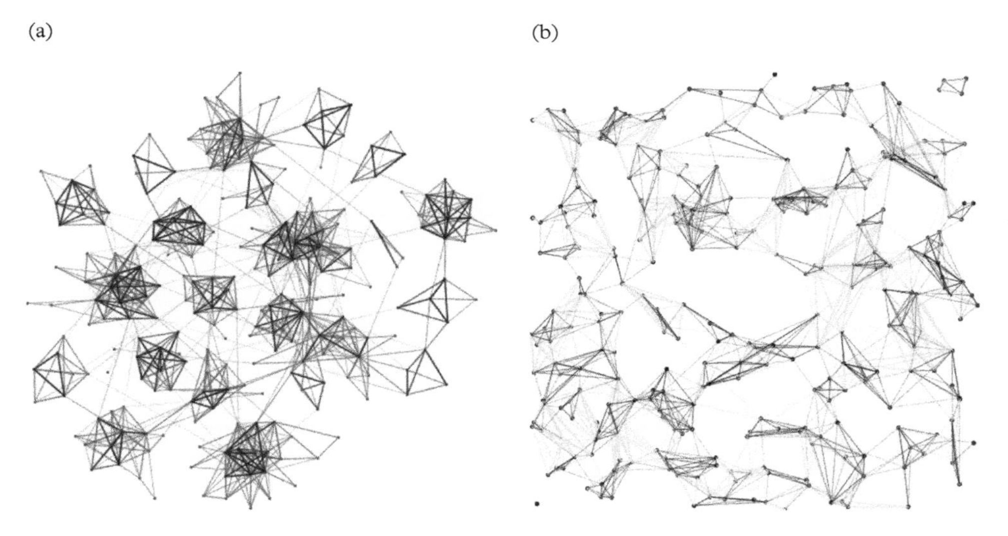


**Figure 5.2** *Network snapshots for (a) the single-layer WSN model and (b) the geographic multilayer WSN model for $\alpha = 6$. In (b), links in the first and the second layers are shown in thin black lines and thicker gray lines, respectively.*

WSN model. After the networks are constructed in each layer, the aggregate network is created by summing up the edge weights: $w_{ij} = \sum_{k=1}^{L} w_{ij}^k$, where $w_{ij}^k$ is the weight of the link between nodes $i$ and $j$ in the $k$th layer. It is this aggregate network for which we expect the coexistence between the overlapping community structure and the stylized facts already reproduced by the original WSN model. In the following, $N = 50{,}000$, $p_r = 0.0005, p_\Delta = 0.05, p_d = 0.001, \delta = 1$, and $w_0 = 1$ are used. The results are obtained after $25 \times 10^3$ time steps and averaged over 50 realizations.

It turns out that this naive multilayer model did not fulfill our expectations. Figure 5.3 shows the percolation analysis for a single-layer network ($L = 1$) and a double-layer network ($L = 2$) to verify the existence of the Granovetterian structure. These two plots show the results for link removal in ascending and descending orders of the link weights. We define $f_c^a$ ($f_c^d$) as the percolation threshold for the ascending (descending) order, marked by the disappearance of the largest connected component and the peak in the second moment of the component size distribution (also called "susceptibility"). The Granovetterian structure is characterized by a significantly large value of the difference $\Delta f_c = f_c^d - f_c^a$ between the two threshold values; in the descending order, the network gets fragmented earlier.

For $L = 1$, we get $\Delta f_c \approx 0.35$, while, for $L = 2$, the percolation threshold for the ascending order $f_c^a$ is approximately the same as that for the descending order $f_c^d$,

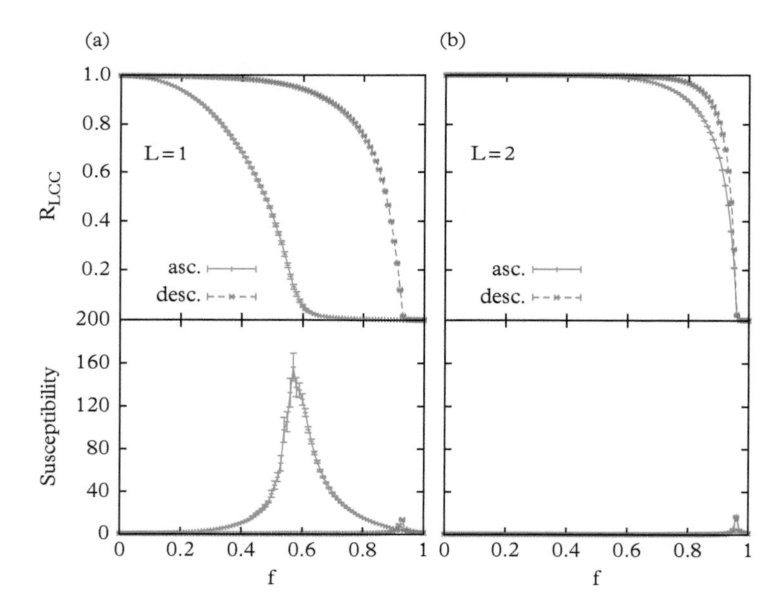


**Figure 5.3** *Link percolation analysis for L = 1 (left) and L = 2 (right). The upper panels show the relative size of the largest connected component, $R_{LCC}$, as a function of the fraction of the removed links f. The lower panels show the susceptibility $\chi$. Solid (dashed) lines correspond to the case when links are removed in ascending (descending) order of the link weights. The figure was taken from Ref. [202].*

leading to $\Delta f_c \approx 0$. This indicates that the introduction of the second layer destroys the Granovetterian structure. The percolation threshold agrees well with that of an Erdős-Rényi (ER) random network having the same average degree $\langle k \rangle$ as the simulated model: $f_c = 1 - 1/\langle k \rangle$, with the measured $\langle k \rangle = 21.9$. This observation shows that combining two independent layers of the original single-layer WSN model already leads to a high level of randomization in the aggregate model. Since strong links, which are intra-community links in the first layer, bridge the communities in the second layer, the difference between the roles of the links with different strength of weights disappears. The simulation results indicate that the empirical networks in different communication contexts cannot be independent. Hence, interlayer correlations play a pivotal role when modeling the multiplex structure of the social network.

### 5.2.2 A geographic multilayer WSN model

The above results show that correlations between layers are essential in order to have $\Delta f_c$ for a multilayer model significantly different from zero, that is, to reproduce the Granovetterian structure in a multiplex setting. Previous studies have reported that there are strong geographic constraints on social network groups, even in the era of the Internet [216], and this is reflected in the CDR data [159, 166, 90]. For example, intercity communication intensity is inversely proportional to the square of the Euclidean distance, which is reminiscent of the law of gravity [159, 166, 173].

Motivated by these observations, we consider a model embedded in a two-dimensional geographic space. Nodes are given fixed position in the unit square with a periodic boundary condition that is shared by all layers. The probability of new links being created by the GA process is proportional to $r_{ij}^{-\alpha}$, where $r_{ij}$ is a distance between nodes $i$ and $j$ and where $\alpha$ is a new parameter controlling the dependence on geographic distance, as in Refs [155, 69]. When $\alpha = 0$, this probability is independent of the geographic distance; thus, the model is equivalent to the uncorrelated multilayer model we presented in the previous section. When $\alpha$ is larger, the nodes will tend to be connected with nodes that are geographically closer. Since the GA process creates links between non-connected nodes, we choose the following normalized connection probabilities in GA:

$$p_{ij} = \frac{r_{ij}^{-\alpha}}{\sum_{k \in S_i} r_{ik}^{-\alpha}}, \tag{5.1}$$

where $S_i$ is the set of the nodes not connected to the node $i$. The other rules, LA and ND, are kept the same as in the original WSN model. Because the network for larger $\alpha$ has a smaller average degree, we used a larger value of $p_r = 0.002$ in the following in order to keep the average degree comparable with the results for the non-geographic model.

The link percolation analysis was conducted for the geographic model using various $\alpha$ values. When $\alpha$ is close to zero, the model does not show the Granovetterian structure, that is, $\Delta f_c \approx 0$. Since the network in this case has no significant geographic effect, the model is essentially equivalent to the naive multilayer model. As $\alpha$ gets larger, $\Delta f_c$ starts to become larger than zero. The dependence of $f_c$ on $\alpha$ is shown in Figure 5.4. The

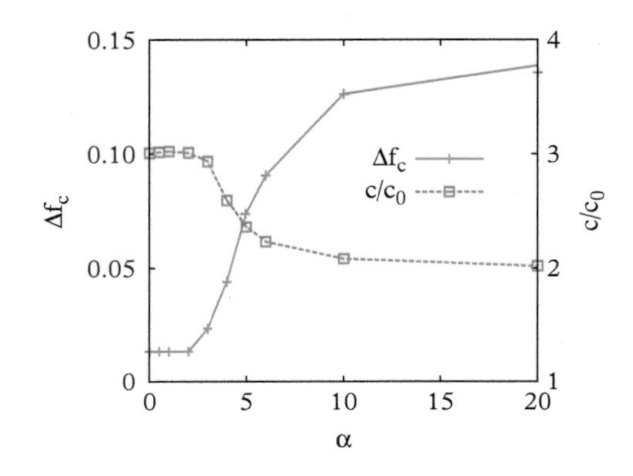


**Figure 5.4** *Characteristic quantities for the geographic multilayer WSN model. The difference $\Delta f_c$ between the percolation thresholds is shown as a function of $\alpha$. The ratio $c/c_0$ is also shown, where $c$ ($c_0$) is the number of communities a node belongs to for the multilayer (single-layer) model. The figure was taken from Ref. [202].*

difference $\Delta f_c$ becomes larger with increasing $\alpha$ and seems to get saturated around 0.15. As shown in Figure 5.4, the network for $\alpha = 6$ exhibits a Granovetterian structure due to $f_c^a$ and $f_c^d$ being significantly different, with $\Delta f_c \approx 0.1$.

A small sample of networks ($N = 300$) for $\alpha = 6$ is shown in Figure 5.2(b). The network for a large $\alpha$ clearly shows a community structure. Due to the correlations between the layers, the network has overlapping communities while maintaining the Granovetterian weight–topology relationship.

The geographic extension of the model produces a region of $\alpha$, where a multilayer Granovetterian structure exists. Now we have to check whether our construction would also lead to the enhancement of community overlap. We analyzed the aggregate networks, using the method described by Ahn *et al.* [3] to calculate $c/c_0$. Here, $c$ ($c_0$) denotes the average number of communities a node belongs to, for the multilayer model (for the corresponding single-layer model). If the ratio $c/c_0$ is larger than 1, nodes have significant amount of overlapping communities due to enhancement by the multilayer structure. Figure 5.4 shows the dependence of this quantity on $\alpha$. The ratio $c/c_0$ decreases rather rapidly when $\alpha$ increases from 0, down to a value of 2. This means that, for sufficiently large $\alpha$, we have *both* Granovetterian properties and an increase in the number of overlapping communities.

The coexistence of Granovetterian structure and enhanced community overlap requires nontrivial correlations between the layers. For example, we have tested a model where the second layer of the network is constructed by replicating the first layer and then the fraction $p$ of the nodes in the second layer gets shuffled. That is, for each pair of nodes $i$ and $j$, with a probability of $p$, all links $ik$ are exchanged with $jk$ only in the second layer. When $p$ increases from 0 to 1, we find a crossover from the single-layer model to the naive multilayer model, and $\Delta f_0$ changes from a finite positive value to 0 as $p$ approaches 1. We also measured the overlap $c/c_0$ for this model but found that the overlap starts to increase only when the Granovetterian correlation between link weight and topology disappears. There is no region of $p$ where both required properties can simultaneously be observed. Thus, an appropriate introduction of the inter-layer correlation, as shown for the geographic model, is necessary.

## 5.3 Modeling channel selection and sampling bias

In recent years, empirical analysis of the society has speeded up due to an access to immense amount of human-related ICT data [218, 315, 135, 83, 298, 165, 217]. Most of the data shows consistent features, as summarized in Subsection 5.1.1. However, as described in the introduction, data are usually collected from a single communication channel, that is, a single layer of the multiplex depicted in Figure 5.1(b). Consequently, the following question remains: to what extent do the results of a single layer of this multiplex network represent the characteristics of the combined and, thus, full social network?

Of course, the best solution would be to combine data from all single-channel layers. This can be done, for example, for transportation networks [105], but, due to technical, privacy, and legal issues, it has been impossible for social data. Therefore, except for cases of reality mining [82, 139, 285] with relatively small number of participants, we are left with single-layer data resulting from a nontrivial sampling mechanism that introduces a bias, as compared to the complete, aggregate network, which is what we are mainly interested in. In this section, we analyze such a sampling by modeling the channel-selection process.

It is known that, in order to preserve the original statistics of the network, one has to do careful sampling [286, 171], and we cannot expect that the way people select their communication channels will obey these rules. This is perhaps most apparent in the form of the degree distribution, which was found to be a decreasing function in almost all datasets (see Subsection 5.1.1). However, it predicts, contrary to all expectation, that the most probable case is someone has just one friend. On the contrary, one would rather expect that maximum of the distribution would be at a degree of the order of the Dunbar number ($\sim$150) [81]. The question arises, if the degree distribution is distorted so much by sampling, then to what extent can one trust observations of other properties, such as assortativity, when only a single communication channel is analyzed?

In order to answer this question, we devised a simple sampling model motivated by natural concepts of human-communication channel selection and show that, this way, we can reproduce some stylized facts obtained by empirical ICT data analysis, even from random networks [295]. In particular, we show how the expected peaked degree distribution gets transformed into a monotonic behavior.

We focus our analysis on two general quantities, namely, the degree distribution $P(k)$ and the average degree of neighbors $k_{\text{nn}}(k)$ of nodes with degree $k$. In Figure 5.5, empirical results are shown for two datasets: a CDR dataset [143] and a dataset for iWiW, the Hungarian online social network that was closed in 2013 but which, for almost 3 years, hosted more than two-thirds of the population with Internet access in Hungary [173]. Both datasets show similar qualitative features even though they are

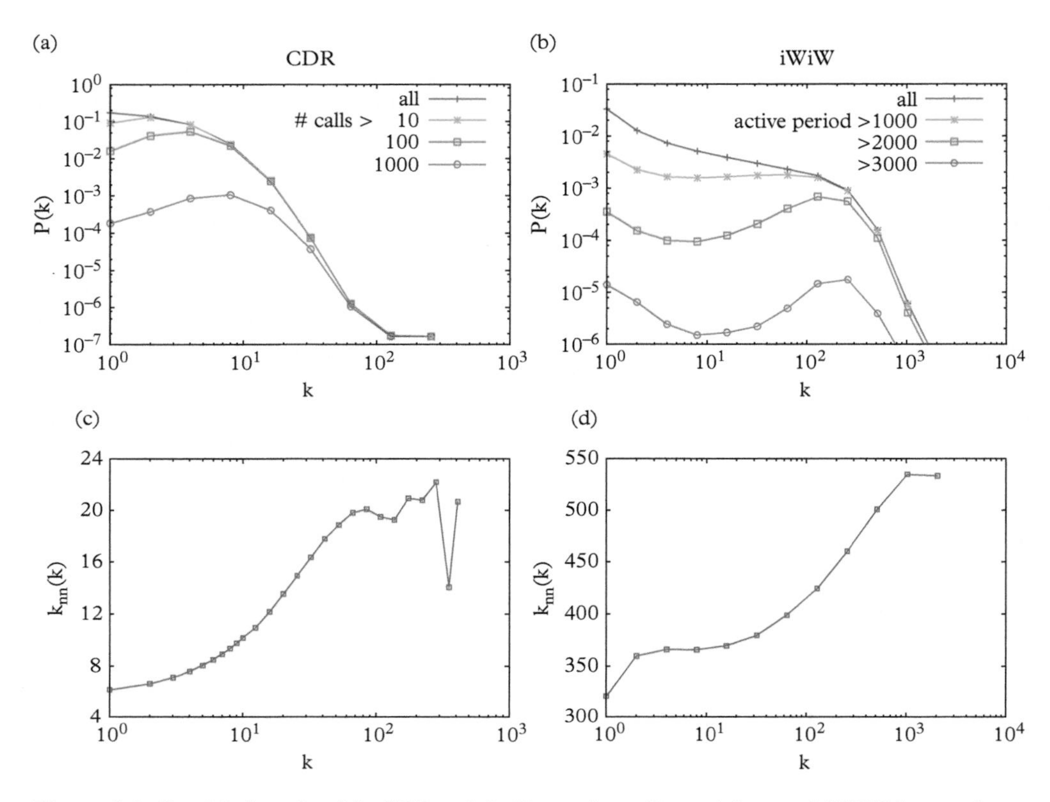


**Figure 5.5** *Empirical results of the CDR and the Hungarian online social network iWiW dataset: degree distributions P(k) of CDR (a) and iWiW (b). We also show* P(k)*-s for nodes with different activity (CDRs) and time spent with the service (iWiW). In the bottom row, the average degree* $k_{nn}$ *of neighbors for nodes with degree k is depicted for the CDR (c) and iWiW (d). Figure is taken from Ref. [295].*

quite different; for example, the average degree is 7.7 for the CDR dataset but 220 for the iWiW dataset. The degree distributions for all nodes in Figure 5.5(a) and (b) are monotonically decreasing functions. Interestingly, if we apply a filter and keep only those users who are sufficiently dedicated to the service, meaning large numbers of calls in the case of the CDR dataset, and longer active periods in the iWiW dataset, the peaked nature of the degree distributions gets brought out. In Figure 5.5(c) and (d), $k_{nn}$ increases with the degree $k$, indicating assortativity. Even the behavior of the second derivative looks similar. It is, however, unclear whether these features reflect the properties of the underlying social network or whether they are due to sampling bias.

We therefore tried to model the process by which people choose communication channels, to see how a surrogate network representing the (unknown) true social network would be transformed. We chose three different networks: a regulat random graph (RR), an ER graph, and a link-deletion version of WSN [201]. All three networks have a peaked degree distribution; RR and ER show no assortative mixing, although WSN does.

When people want to communicate, they have to choose a channel of communication. Naturally, people have diverse preferences and different people may favor different communication channels. However, sticking to someone's favorite does not make sense; for example, writing a message on a chat server to someone who checks his account only weekly is rather meaningless, and so is calling someone who never picks up the muted phone. In order to make the communication successful, one has to resort to the least inconvenient channel for both of them. To make it more quantitative, we assigned an *affinity* to an individual $i$ toward a communication channel $v$, denoting this value $f_i^v$. We assume that the probability of choosing the communication channel $v$ for individuals $i$ and $j$ is proportional to the smaller of the two affinities: $p_{ij} = \min(f_i^v, f_j^v, 1)$. Thus, the probability of a link existing in layer $v$ will also be proportional to $p_{ij}$.

Our model for the sampling effect of a single communication channel is thus defined as follows. Let us consider a surrogate network. Each node is given a randomly chosen affinity $f$ toward this specific communication channel. The affinities are taken from an exponential distribution, reflecting that there are always many more people who put a small amount of effort into using a specific ICT service than those few who are really addicted to it:

$$P(f) = \frac{1}{f_0} e^{-f/f_0}. \tag{5.2}$$

The links in the sampled network are kept with probability

$$p_{ij} = \min(f_i, f_j, 1). \tag{5.3}$$

All nodes which have at least one link are kept for the sampled network.

We tested our sampling model with the following surrogate networks: ER and RR, with an average degree of $k_0 = \langle k \rangle = 150$, and WSN, with $\langle k \rangle \simeq 47.8$.[1] The results are shown in Figure 5.6(a) and (b). Clearly, the originally peaked distribution has become monotonically decreasing by sampling. It is also interesting that the shape of the curve depends only very little on the original degree distribution, as demonstratively shown in Figure 5.6(a). Here, we find the marginal difference between the degree distributions of RR and ER.

We can carry out a similar filtering as before for the single-layer empirical data by selecting nodes dedicated to the channel. The high-affinity nodes show progressively peaked degree distributions in Figure 5.6(c), which indicates that, indeed, the properties of such nodes are closer to the original network than low-affinity ones are.

The sampled degree distribution can be calculated analytically [295], using the fact that the affinities are assigned randomly, so there is no correlation between the affinities of neighboring nodes in the surrogate networks. The degree distribution $Q_{k_0}(k)$ for the RR network with degree $k_0$ is

[1] Here, we used the link-deletion WSN model proposed in Ref. [201]. The parameters used to generate the WSN were $N = 10^4$, $p_\Delta = 0.07$, $p_r = 0.0007$, and $p_{ld} = 0.0015$, and the maximum time step $t_{max} = 50{,}000$.

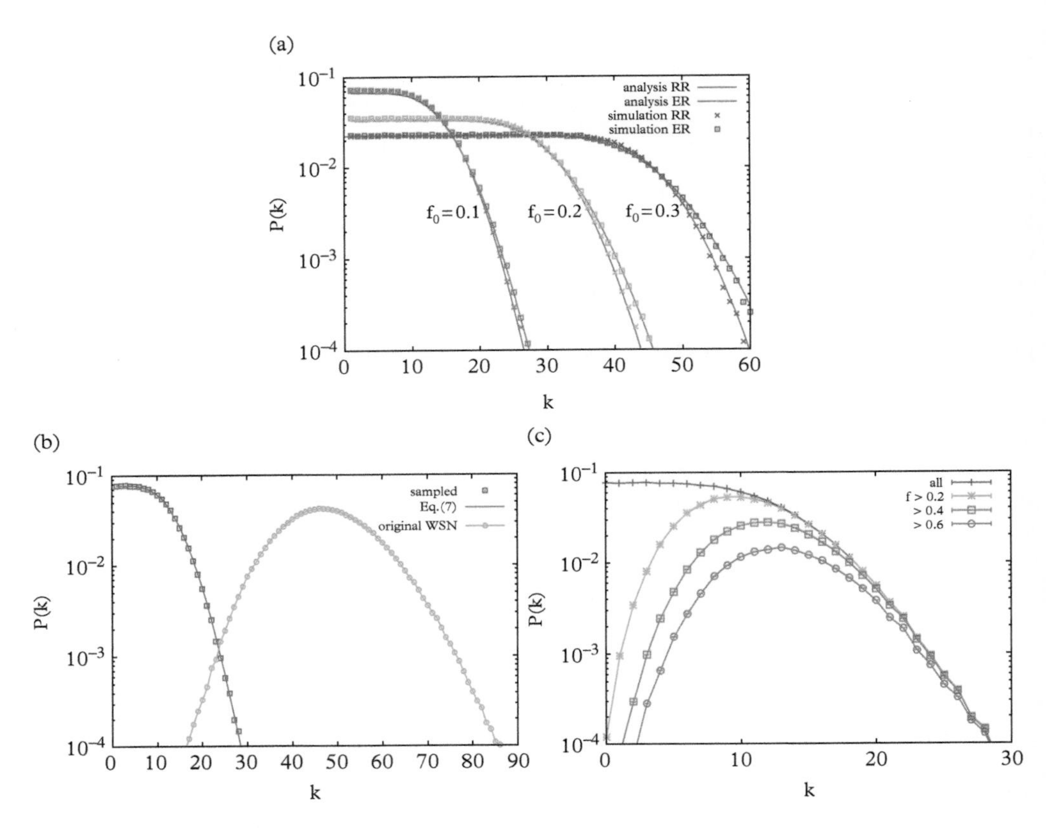


**Figure 5.6** *(a) Degree distributions of sampled networks when RR and ER with $k_0 = \langle k \rangle = 150$ and $N = 10^4$ are used as surrogate networks, using $f_0 = 0.1, 0.2,$ and 0.3. Solid lines denote analytic solutions of Eq. (5.5). (b) Degree distributions of original and sampled networks using the WSN model as surrogate network with $f_0 = 0.3$. The solid line denotes the degree distribution obtained using Eq. (5.5). (c) Degree distribution of the sampled network from WSN, using $f_0 = 0.3$ and those when restricted only for nodes having an affinity above the indicated threshold. Figure is taken from Ref. [295].*

$$Q_{k_0}(k) = \frac{1}{f_0(k_0+1)} I_{\left(\frac{f_0}{1-f_0}\right)}(k+1, k_0-k+1), \tag{5.4}$$

where $I_x(a,b)$ denotes the regularized beta function.

For general degree distributions, for the case of uncorrelated affinities, the degree distribution can be obtained as a weighted sum:

$$P(k) = \sum_{k'=0}^{\infty} P_0(k') Q_{k'}(k). \tag{5.5}$$

When Eq. (5.5) is verified against the numerical data in Figure 5.6(a) and (b), the match is perfect.

We have shown that our channel-selection model reproduces the observed effect, namely, the transformation from a peaked degree distribution to a monotonically decreasing distribution due to sampling. Now we turn our attention to the assortativity as calculated from $k_{nn}(k)$, the average degree of the neighbors for nodes with degree $k$ (see Figure 5.7). The sampled results for both ER and WSN show assortative mixing similar to that found with the empirical data, even though it is known that ER has degree-independent $k_{nn}$. This demonstrates that sampled networks can show similar assortative behaviors, irrespective of their original properties. Again, considering nodes in each given affinity range, we find flat behavior for $k_{nn}(k)$ for those nodes, as in the surrogate networks, as depicted in Figure 5.7(a).

A remark must be made at this point: our channel-selection model as described by Eqs (5.2) and (5.3) is certainly a crude approximation of reality. Therefore, in order to check the robustness of our results, we applied the generalized mean instead of taking the minimum of the affinities for the selection rule:

$$p_{ij} = \left(\frac{f_i^\beta + f_j^\beta}{2}\right)^{1/\beta}, \tag{5.6}$$

with $\beta \to -\infty$ providing the rule of Eq. (5.3) used above. We have shown that we have a decreasing degree distribution in the sampled networks only when $\beta$ is negative and that assortativity is generated for this parameter region even if we use uncorrelated surrogate networks [295].

Our simple model of communication channel selection shows that the sampled network resulting from this selection mechanism may seriously distort the properties of

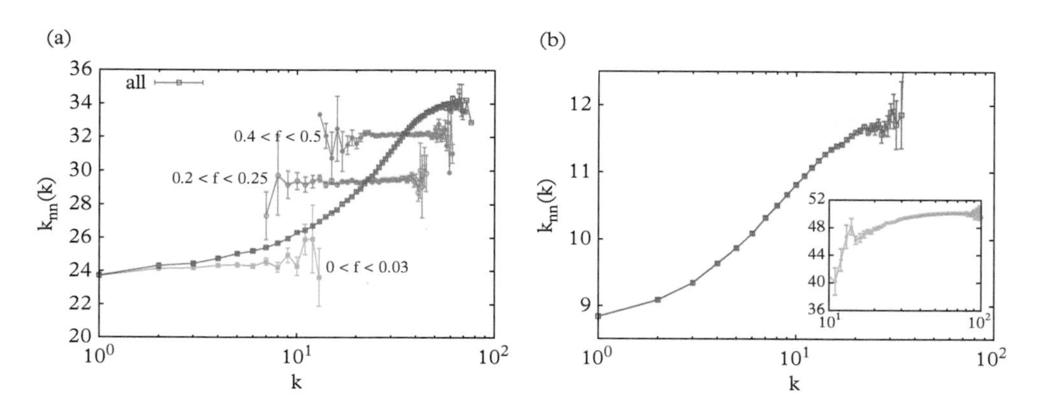


**Figure 5.7** *(a) Average degrees of neighboring nodes as a function of the node degree for sampled network using ER graphs as surrogate networks for all nodes, and for a range of node affinity. (b) Assortativity of the sampled network as a function of the degree when WSN is used as a surrogate network. The inset shows the case for the surrogate network. Figure is taken from Ref. [295].*

the original network. As most of the nodes have small affinities, their social network will be poorly represented in a given ICT network. The nodes with high degree in the sampled network are not necessarily the ones that had the most contacts in the original network but rather are the ones with high affinity toward this particular service. This distorts the network in such a way that new features can be observed. The sampling model presented here has so strong influence on the network properties that it may completely hide the original ones and shows the biased properties instead. This emphasizes that single-channel empirical data should be handled with care.

Our study also demonstrated that we may get some insight into the real structure of the original network properties if the analysis is restricted to a subset of well-embedded nodes from the sampled network. In our calculations, we used the affinity of the nodes as a measure of this embeddedness but our results with CDR and SNS data indicate that activity or time spent with the service may also be used for this purpose.

## 5.4 Summary and discussion

In this chapter, we discussed how society can be considered as a multiplex with respect to the nature of the links reflecting the contexts of the interactions between persons in the society (generative aspect) or from the point of view of the communication channels analyzed (data collection aspect; see Figure 5.1). We have shown how both Granovetterian structure and community overlap can be maintained in a multiplex model. In order to do so, we started from a single-layer WSN model [162] and generalized it to a multiplex. However, a naive introduction of multiple layers to a single-layer WSN model breaks the Granovetter-type weight–topology relation, so instead we introduced geographic correlations between the layers.

Our results have several implications. First, we have shown the importance of correlations between layers. Moreover, it seems that specifically geographic correlations may play a key role in maintaining the stylized facts in a multiplex weighted network. It is worth noting that the peculiar role of geographic correlations was observed for such networks earlier [155, 19], including interdependent networks [175]. We mention that communities may organize themselves along various diverse but common attributes such as shared working places, classes at universities, joint interests, for example, in sport, and residential districts. However, all these have some geographic aspect. In fact, even in the digital era, distance is not “dead” [118, 173], contrary to some earlier speculations [53]. Of course, the consideration of further realistic correlations should improve the model.

The other multiplex aspect of society is related to the different communication channels that can be used. Due to the fact that our data analytics mostly relies on observations from a single communication/interaction channel, the question arises, to what extent can ICT data tell us about the structure of the entire social network of people, as all such types of data are incomplete and capture only a part of the whole plethora of social relationships? This is closely related to the important question of channel selection, which we have attempted to model here.

While ICT services are diverse, we nevertheless observe some common features, for example, they all display an overall decreasing degree distribution, which cannot be true for the entire social network and hence should be attributed to the sampling. To investigate the effect of sampling by single-channel selection, we modeled how people use ICT communication services. Using simple assumptions, we were able to reproduce robustly the stylized facts of the ICT data, namely, the decreasing degree distributions and assortative mixing, even when they were absent in the original surrogate networks. Our results resolve the long-lasting contradiction between the observed and expected shapes of the degree distributions. Moreover, they call attention to the danger of misinterpreting observations from single-channel data for the entire social interaction network. At the same time, we have also shown that there is a subset of users with high activity, that is, users who put much effort into the given ICT service, whose characteristics are at least qualitatively in accordance with those of the original surrogate network. This feature hints toward a possible resolution of the problem of the sampling bias.

Our results rely on the model of channel selection as expressed by Eqs (5.2) and (5.3) and their generalizations. We have shown that there is a class of rules that result in the universally observed single-channel properties of monotonic degree distribution and assortative mixing. Such class of rules are similar to the minimum rule shown in Eq. (5.3), that is, a person does not select a communication channel with a friend who does not like that channel, even if that is the first person's favorite.

It should be mentioned at this point that we consider our channel-selection model to be only a first step in this very interesting problem. Clearly, the assumption of uncorrelated affinities should be revised. Homophily, one of the most important factors in tie formation [190], implies that there are strong similarities in the affinities of neighbors. Also, node properties, such as age and gender, should influence affinity values. These features may generate higher-order correlations, making it possible to deal with the effect of sampling on clustering and communities.

The task of large-scale modeling of society has just started. The activity in this field is increasing, and several attempts at such models have already been published. Here, we focused on our own contributions but we could also have mentioned, for example, the recent work by Battiston *et al.* on multilayer modeling of a given community structure [26] or the very interesting model of a virtual multilayer society by Klimek *et al.* [152]. Although these models are very simplified, we believe that they contribute to the understanding of social structures. Moreover, adequate models enable us to investigate the impact of societal structure on dynamic phenomena such as spreading. Future work in such directions is expected.

## ACKNOWLEDGMENTS

J. K. acknowledges support from EU Grant No. FP7 317532 (MULTIPLEX). J. T. gives thanks for financial support from the Aalto AScI internship program. Y. M. expresses appreciation for the hospitality received Aalto University and acknowledges support from CREST, JST. H.-H. J. acknowledges financial support from the Basic Science

Research Program through the National Research Foundation of Korea (NRF) grant funded by the Ministry of Education (2015R1D1A1A01058958) and the framework of international cooperation program managed by the National Research Foundation of Korea (NRF-2016K2A9A2A08003695). This project was partly supported by JSPS and NRF under the Japan–Korea Scientific Cooperation Program. Partial support by OTKA, K112713, is also acknowledged. The systematic simulations in this study were assisted by OACIS [203]. K. K. acknowledges support from the Academy of Finland's COSDYN project (No. 276439) and the EU Horizon 2020 FET Open RIA 662725 project IBSEN.

# 6

# Data Summaries and Representations: Definitions and Practical Use

**Alain Barrat**[1,2] **and Ciro Cattuto**[2]

[1]Aix Marseille Univ, Université de Toulon, CNRS, CPT, Marseille, France
[2]Data Science Laboratory, ISI Foundation, Torino, Italy

## 6.1 Introduction

Complex networked data have become available in a variety of contexts, describing a variety of systems with growing abundance of details, such as, for instance, the multiple natures of links between individuals in social networks, or the temporal evolution of these links. The availability of such rich datasets describing, for instance, the behavior and interactions of individuals or socioeconomic entities is bringing forth both new opportunities and challenges. Data comes from heterogeneous sources, at different scales and resolutions and with variable amounts of details or metadata and sometimes temporal resolution. Data alone, however, even in huge quantities, does not easily transform into knowledge or predictive models. The richness, level of detail, and diversity of datasets raise crucial challenges concerning data analysis, representation, and interpretation; the extraction of structures from data; and the practical use of data, be it to compare different systems, explore their temporal evolution, or integrate data into data-driven models of interest in contexts such as epidemiology or computational social sciences.

Data thus need to be summarized and represented in simplified forms. To this aim, one needs to understand which characteristics of any dataset under investigation are crucial to retain and which ones, on the other hand, are too specific to be of general interest. A data representation encapsulates relevant information while discarding unnecessary details. Data representations can be more or less summarized or coarse-grained with respect to the original data. For any dataset, one can define a number of representations retaining different amounts of information about the characteristics of the data, and

Barrat, A. and Cattuto, C., "Data Summaries and Representations: Definitions and Practical Use" in *Multiplex and Multilevel Networks*, edited by Battiston, S., Caldarelli, G., and Garas, A. © Oxford University Press 2019.
DOI: 10.1093/oso/9780198809456.003.0006

the choice of the most useful representation will depend on the specific goal and its specific use.

In this chapter, we focus on temporal networks. We describe several commonly used data summaries and levels of representation of temporal networks, as well as novel data representations that have been developed through the MULTIPLEX project. We focus in particular on the case of temporal networks of contacts between individuals and show in a series of concrete use cases how different representations can be used to characterize and compare data or to feed data-driven models of epidemic spreading processes.

## 6.2 Data and representations of data

### 6.2.1 Datasets, common summaries, and representations

Let us consider a dataset describing a temporal network, that is, a set of nodes representing, for instance, individuals, and the links that appear and disappear between these nodes. This is the case for the numerous datasets of face-to-face contacts between individuals that have been collected by the SocioPatterns collaborations and other groups in a number of countries and contexts, including schools, hospitals, scientific conferences, museums, and so on. Each such dataset is typically of the following detailed nature: for each pair of individuals $i$ and $j$, the dataset contains a list of $\ell$ "events," that is, successive time intervals $((t_{ij}^{(s,1)}, t_{ij}^{(e,1)}), (t_{ij}^{(s,2)}, t_{ij}^{(e,2)}), \ldots, (t_{ij}^{(s,\ell)}, t_{ij}^{(e,\ell)}))$ during which $i$ and $j$ were detected to be in close-range face-to-face proximity, where $t_{ij}^{(s,a)}$ refers to the starting time, and $t_{ij}^{(e,a)}$ to the ending time of the time interval number $a$. Note that, in a number of temporal networks, such as networks of communications between individuals, the durations of the events are neglected, so that each event is composed only of one timestamp. While this representation contains all the available data, and hence retains all the available information, it entails some disadvantages. On the one hand, the visualization of a time-evolving network is challenging, making it difficult to grasp its structures and features. On the other hand, a specific dataset is often unique and differs from other similar datasets describing, for instance, the same system at another time, or similar systems. Examining the full dataset without using the lens of coarse-grained representations or summaries can then inhibit the search for commonalities and robust patterns or properties. For instance, the detailed face-to-face contacts that occur in a specific school on a specific day are certainly unique but bear some important similarities with the contacts of another day in the same school, even if they do not repeat in the same way every day. Using summaries and representations makes it possible to highlight similarities and pinpoint important differences between similar datasets.

#### *6.2.1.1 Statistics*

The first type of data representation that is customarily used consists in building statistics for several quantities of interest. For instance, the temporal evolution of the number of events per unit time can inform us about circadian patterns in the data and about the possible recurrence of moments of high and low activity. The evolution of the number of events involving each individual, as well as of the number of other individuals with whom an event is shared, can also reveal interesting patterns.

Moreover, the list of contact time intervals yields for each pair of individuals $i$ and $j$ a list of contact durations $(\Delta t_{ij}^{(1)}, \ldots, \Delta t_{ij}^{(\ell)})$, with $\Delta t_{ij}^{(a)} = t_{ij}^{(e,a)} - t_{ij}^{(s,a)}$ for $a = 1, \ldots, \ell$. The distributions of these contact durations, as well as of the time intervals between contact events, have been found to be broad in many datasets: most contact durations and intervals between successive contacts are very short, but very long durations are also observed, and no characteristic timescale emerges. This bursty behavior is a well-known feature of human dynamics and has been observed in a variety of systems driven by human actions, with important consequences on processes unfolding on temporal networks.

While these properties are, of course, of interest, they certainly do not reveal enough to fully characterize a dataset. Strikingly, a number of temporal network characteristics are, in fact, defined through the use of another, more coarse-grained representation that is widely used and sometimes implicitly considered: the temporally aggregated network.

#### *6.2.1.2 Aggregated networks*

The sequence of events between the nodes of a temporal network during a given time window defines an aggregated network, which is a static summary of the temporal network. Taking once again the example of a temporal contact network, each node of the aggregated network is an individual, and a link between two nodes $i$ and $j$ denotes the fact that the corresponding individuals have been in contact at least once during the time window under consideration. The bare structure of this graph encodes information on the overall topological structure of the temporal network, but not on its temporal properties. In order to retain some temporal information, it is customary to summarize the temporal activity of individual edges $i$–$j$ by suitably defined weights for the edges. Several notions of weight $w_{ij}$ for the edge $i$–$j$ can be defined on the basis of the list of contact durations, yielding weighted contact networks that describe different aspects of the empirical sequence of contacts:

- edge presence: $w_{ij}^p$ measures the contact occurrence (the superscript $p$ stands for "presence"), with $w_{ij}^p = 1$ if at least one contact between $i$ and $j$ has been established, and 0 otherwise;
- frequency of occurrence: the frequency $w_{ij}^n = l$ indicates how many distinct contact events have been registered between $i$ and $j$, disregarding the length of each contact (the superscript $n$ is for "number");
- cumulative time in contact: the cumulative duration of the contact $w_{ij}^t = \sum_a \Delta t_{ij}^{(a)}$ gives the sum of the durations of all contacts established between $i$ and $j$.

The time window considered for aggregation can range from the finest time resolution of the data up to the entire duration of the dataset. In many contexts, it is natural to consider a specific temporal aggregation scale (i.e., daily), but different aggregation levels typically provide complementary views of the network dynamics at different scales.

The aggregated network representation carries both advantages and limitations. An important feature of aggregated networks is their static nature: this makes them amenable to the usual characterization tools of network analysis and visualization, such as degree distribution, clustering, assortativity, and so on. A comparison of their structures across contexts can reveal important information about the contact patterns of the population, as we will see later on in concrete examples. Moreover, the assignment of weights to links makes it possible to keep track of important characteristics such as the heterogeneity of the number and duration of contact events between different pairs of individuals, and of higher-order correlations between the number and duration of events between individuals. Aggregation on successive time windows can also shed light on the temporal evolution and possible stability of the system under scrutiny. This can be done at different levels of detail, using a comparison of (degree, weight, strength) distributions measured on different time windows, or a measure of a similarity, such as a Jaccard coefficient between successive aggregated networks. Such a measure can even be used to automatically detect relevant timescales in temporal networks [70]. At a finer resolution, we can investigate the similarity between the neighborhoods of a given node in contact networks aggregated over different periods. For instance, for daily aggregated networks, the similarity between the neighborhoods of an individual $i$ in contact networks measured on 2 different days, which are denoted 1 and 2, respectively, can be quantified through the cosine similarity $\sigma^{1,2}(i) = \sum_j w_{ij,1} w_{ij,2} / \sqrt{\sum_j w_{ij,1}^2 \sum_j w_{ij,2}^2}$, where $w_{ij,d}$ is the weight of the link between $i$ and $j$ in the contact network of day $d$, that is, the cumulative duration of the contacts between $i$ and $j$ occurring on day $d$. The cosine similarity takes values between 0 and 1: it is equal to 0 if $i$ had contact with different individuals in the 2 days considered, and to 1 if $i$ had contact with the same person on both days, with proportional durations.

Obviously, aggregated networks also have limitations. The main one stems from the fact that they do not provide information about the order of events. Different temporal networks with different histories can thus give rise to the same aggregated weighted network, as shown, for example, in Figure 6.1. This can turn out to be crucial when dealing with processes on temporal networks: if, for instance, $A$ and $B$ come into contact before $B$ and $C$ do, $A$ can transmit information to $C$ through $B$ while, if the order of contacts is reversed, this propagation path does not exist. A static representation does not distinguish between these possibilities and therefore overestimates the existence of paths between nodes. As a result, aggregated networks can yield misleading results about the relative importance of nodes in the network (as measured, e.g., by their centralities) [135, 230, 257].

#### *6.2.1.3 Contact matrices*

When the population described by the data at hand can be divided into groups, such as, for instance, age groups, classes in a school, departments in companies, and so on, it is customary to describe the contact networks between specific groups of individuals by using contact matrices, which contain very coarse summaries of the data but highlight

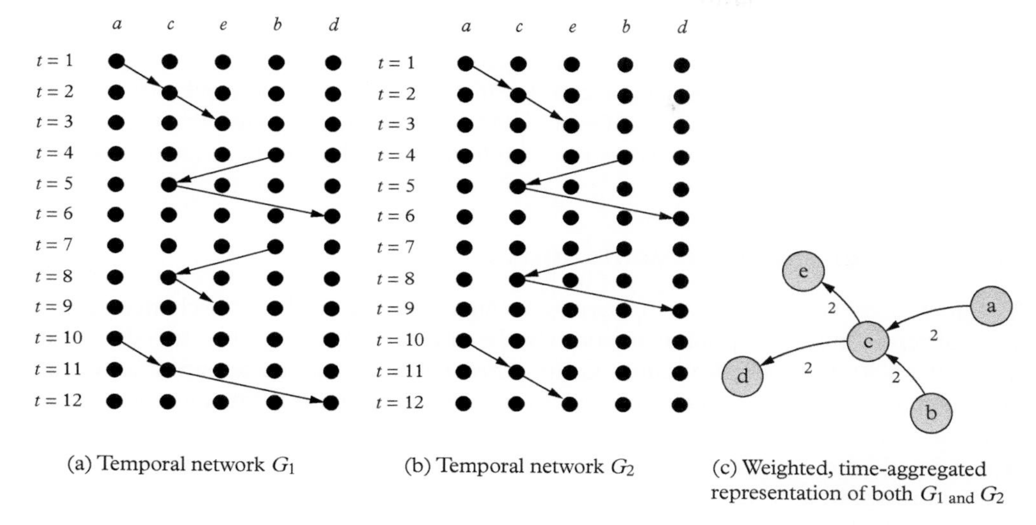


(a) Temporal network $G_1$

(b) Temporal network $G_2$

(c) Weighted, time-aggregated representation of both $G_1$ and $G_2$

**Figure 6.1** *Time-unfolded and weighted static, time-aggregated representation of two temporal networks $G_1$ and $G_2$; two different temporal networks can yield the same aggregated weighted network. From Ref. [257], with kind permission of The European Physical Journal (EPJ).*

the mixing patterns between these groups. If the population is divided into $n$ groups and if we denote the number of individuals in group $X$ by $n_X$, one usually considers the following quantities, aggregated over each time window of interest:

- the total number of contacts between individuals of class $X$ with individuals of group $Y$: $N_{XY} = \sum_{i \in X, j \in Y} w^n_{ij}$ (for $X = Y$, we have $N_{XX} = \frac{1}{2} \sum_{i,j \in X} w^n_{ij}$),
- the average number of contacts of an individual of group $X$ with individuals of group $Y$: $n_{XY} = N_{XY}/n_X$,
- the total time spent in contact between individuals of group $X$ with individuals of group $Y$: $W_{XY} = \sum_{i \in X, j \in Y} w^t_{ij}$ (for $X = Y$, we have $W_{XX} = \frac{1}{2} \sum_{i,j \in X} w^t_{ij}$),
- the average time spent by an individual of group $X$ in contact with individuals of group $Y$: $w_{XY} = W_{XY}/n_X$.

Contact matrices thus contain an even more coarse-grained representation of the data than aggregated networks do: the differences between individuals of a group are neglected, as is the specific structure of the contact network, since only averages are retained. They are, however, widely used to inform data-driven models of spreading processes in the epidemiology of infectious diseases: when dealing with diseases that affect persons of different ages in different ways, for instance, it can indeed be crucial to take into account the fact that children have more contacts among them than adults do, with therefore a higher propagation risk. The specific structure of the contact networks of children and adults might, on the other hand, be less relevant. In this respect, the simplicity of contact matrix representations is appealing. Moreover, even such a simple representation can carry interesting information about the temporal stability of the mixing patterns between groups: one can, for instance, consider the similarity between two contact matrices measured in different time windows, which can inform us about the differences between the mixing patterns of children and adults during schooldays versus weekends or vacations.

## 6.2.2 Novel data representations

The above description of data representations highlights the need for novel intermediate ways of representing temporal networks. On the one hand, keeping too much detail can limit the ability to generalize data. On the other hand, aggregated temporal networks, even if weighted, do not take into account enough temporal information to correctly rank nodes by their importance, and they overestimate the existence of paths between nodes; in addition, contact matrices do not take into account the heterogeneity of links and nodes within a group and discard any structure, in particular, the fact that not all pairs of nodes are linked.

Within the MULTIPLEX project, two complementary directions were followed in the development of novel frameworks for the representation of temporal network data. Importantly, both correspond to static representations and hence are much easier to deal with than temporal ones but retain more temporal information than the representations discussed above.

### *6.2.2.1 Higher-order aggregated networks*

An important first issue, given a temporally resolved dataset, is determining in which measure an aggregated representation gives inaccurate information on the real data. To this aim, the betweenness preference was introduced in Ref. [230]: it quantifies to what extent paths existing in time-aggregated representations of temporal networks are actually realizable in the time-resolved data. In other words, measuring betweenness preference in empirical temporal networks makes it possible to determine whether the corresponding aggregated representations lose too much information or can be used for the simulation of dynamical processes unfolding on these networks.

As discussed above, static representations of temporal data are, however, of great interest. To go forward in this perspective, several authors have therefore introduced new higher-order time-aggregated representations of temporal networks [243, 258] that take into account non-Markovian effects and thus preserve causality. An example is given in Figure 6.2 for second-order aggregation: each second-order node represents an edge in the first-order aggregate network $G^{(1)}$; second-order edges are given by all pairs $(e_1, e_2)$ of directed edges of type $e_1 = (a, b)$ and $e_2 = (b, c)$ in $G^{(1)}$, that is, by all possible paths of length 2 in the first-order aggregate network. Second-order edges are, moreover, weighted, and the weights $w^{(2)}(e_1, e_2)$ can be defined as the relative frequency of time-respecting paths $(a, b; t_1) \rightarrow (b, c; t_2)$, with $t_2 > t_1$ of length 2 in the temporal network.

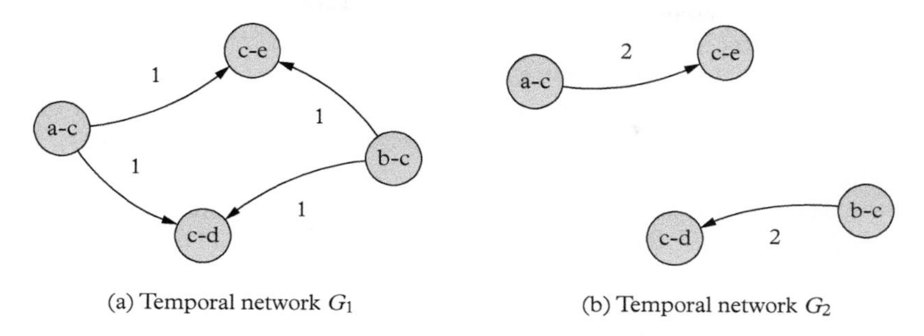


(a) Temporal network $G_1$

(b) Temporal network $G_2$

**Figure 6.2** *Second-order aggregate networks corresponding to the two temporal networks shown in Figure 6.1. From [257], with kind permission of The European Physical Journal (EPJ).*

Representing data in this way, instead of simply aggregating temporally, makes it possible to keep more information on relevant temporal correlations. For instance, using random walks on such a representation preserves the statistics of temporal paths of length 2, that is, of correlations that can be crucial when simulating dynamical processes on top of temporal networks, and to create a novel, causality-preserving, null model of temporal networks. This representation has made it possible to show, for instance, that non-Markovian characteristics of temporal networks can either enforce or mitigate the influence of topological properties on dynamical processes. As such, they constitute an important additional dimension of complexity that needs to be taken into account when studying time-varying network topologies [258]. Moreover, the path-based centralities defined in such higher-order aggregate networks give a much better evaluation of the ranking of nodes' importance in the temporal network than when the usual aggregated networks are used [257].

### *6.2.2.2 Contact matrices of distributions*

Contact matrices of distributions (CMDs) extend the usual concept of contact matrices to tackle their shortcomings while maintaining a high level of summarization of complex temporal data. In practice, the matrix defined in [181] has as the entry for groups X and Y the distribution of aggregated contact durations between all pairs of individuals x (from group X) and y (from group Y), where the distribution includes the fraction of such pairs that do not have any contact. Alternatively, it is possible to fit all the distributions to a common functional form (in [181], a negative binomial distribution) and to consider as entries of the matrix the parameters of the fits. Like standard contact matrices, the CMD is not an individual-based representation: it does not retain the detailed structure of the empirical contact network and contains only a summary of all interactions between individuals of various groups. However, in contrast with the usual contact matrices, it accounts for the broad fluctuations of contact durations as well as for the potentially high fraction of missing links across groups of individuals (it does not assume that all individuals have been in contact).

It is important to note that the CMD defines, in fact, a representation *framework* that can be extended and refined in various ways. For instance, while the CMD in Ref. [181] contains only the distribution of daily aggregated contact durations, one can retain as the entry of the matrix the more detailed distributions of (i) durations of single contacts (ii) intercontact durations, and (iii) number of contacts between pairs of individuals, as well as the density of links between groups. Keeping this information allows one to build realistic timelines of contacts between individuals that respect these statistics, as we will see in the next section [113]. Moreover, matrix entries could also retain clustering coefficients of the aggregated networks, their assortativity properties, or other temporal or structural properties considered relevant.

### 6.2.3 Detecting mesoscopic structures

The representations discussed above are based on aggregation over time or over node attributes, projecting away many specificities, structures, and correlations of the original data. Depending on the problem at hand, these aggregated representations may overlook or confound important structural features of the network. For example, a node may belong to different communities at different points in time: aggregating the network over time will artificially merge the communities and create a cluster that does not represent the network at any point in time. Groups of nodes with similar activity patterns over time can also exist: for instance, in environments such as schools, the interactions that are possible at a given time are driven and constrained by an externally imposed schedule of social activities (e.g., class and lunch breaks). In this case, temporal aggregation of the network may retain the topology of interactions but loses the information on correlated activity patterns, which may play an important role for, for example, epidemic processes unfolding over the temporal network [111]. In general, correlated topological and temporal features of the network may give rise to structures that are neither local features of individual nodes or edges nor global structures and are hence often called "mesostructures." Detecting such mesoscopic structures in high-resolution social network data is an outstanding challenge that goes beyond the extension of community detection techniques to temporal networks.

In this perspective, a promising approach uses a tensor representation of the temporal network: one starts by representing the temporal network as a time-ordered sequence of adjacency matrices, each describing the state of the network at a discrete point in time. The adjacency matrices are then combined into a three-way tensor that encodes all the information about the temporal network and has been recognized as a convenient representation both for multilayer networks and for temporal networks. As shown in Ref. [110], nonnegative tensor factorization techniques, which have shown their relevance in the field of machine learning, can then be used to extract nontrivial structures and represent such complex data as a sum of simpler terms that can be more easily interpreted. Interestingly, some of these structures correspond to so-called communities in static networks, but others entail a complex interplay of activity and structural patterns that could not be found by usual community detection tools. This opens the door to representing information-rich complex data in simpler, human-readable ways and also to investigations on how each simpler structure impacts dynamical processes unfolding on these data, as we will discuss below.

## 6.3 Putting the data representations to concrete use

### 6.3.1 Comparing datasets

In the task of comparing datasets, to assess, for instance, the robustness of stylized facts concerning the system(s) of interest observed at different moments or under different conditions, even very simple and coarse-grained representations are extremely valuable. For instance, statistical distributions of contact durations and their functional shapes can be compared in datasets describing contacts between individuals collected in different contexts [284, 18, 99, 114, 189, 300]: as shown in Figure 6.3, the distributions of contact durations are very broad, and extremely similar for very different contexts, populations, activity timelines, and deployment conditions. The broadness of these distributions, as well as their robustness, imply two important facts for modelers, in particular when dealing with processes depending on contact durations between individuals, such as epidemic spreading. First, the broadness of the distributions means that taking into account only average contact durations and assuming that all contacts are equivalent might be too coarse a representation of the reality. Indeed, different contacts might yield very different transmission probabilities: many contacts are very short and correspond to a small transmission probability, but some are much longer than others and could therefore play a crucial role in disease dynamics, Second, the robustness of the distributions found in different contexts means that these distributions can be assumed to depend negligibly on the specifics of the situation being modeled and thus directly plugged into the models.

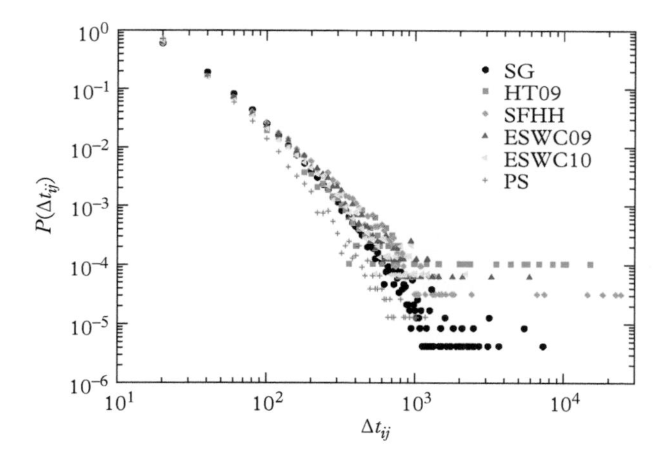


**Figure 6.3** *Distributions of the face-to-face contact durations measured in different environments ranging from a museum (SG) to a school (PS) and several scientific conferences.*

Longitudinal studies can also be carried out at the level of coarse-grained summaries. For instance, the distributions of contact or intercontact durations have been shown to be robust when measured in different time windows, revealing a statistical stationarity in an otherwise non-stationary signal. Activity timelines giving, for example, the number of contacts between individuals of different groups can turn out to be very stable from one day to the next, as, for instance, in hospitals or schools [300, 99, 189], or to have a more casual character, as in offices [114], giving hints on the impact of organizational details on contacts. Contact matrices giving the average number or duration of contacts between individuals of different categories have also revealed an interesting robustness of contact patterns in a high school across different timescales: these contact matrices, computed for the same classes in different days or even in different years, are extremely similar [99, 189], showing, for instance, that temporally limited datasets can already yield important information on students' mixing patterns that remain relevant on long timescales. Finally, contact matrices built from data coming from different sources, namely, wearable sensors and contact diaries, have also revealed a strong similarity, despite raw datasets differing both qualitatively and quantitatively [189], as shown in Figure 6.4.

Notably, both the robustness of distributions of contact durations and that of contact matrices imply the robustness of CMDs, a fact that we will exploit below.

The study of aggregated networks obviously sheds more light on the comparison of datasets, not only by providing additional statistical characteristics such as the distribution of the number of neighbors, the distribution of the cumulative contact durations, and the correlations between nodes' properties, such as degree and strength,

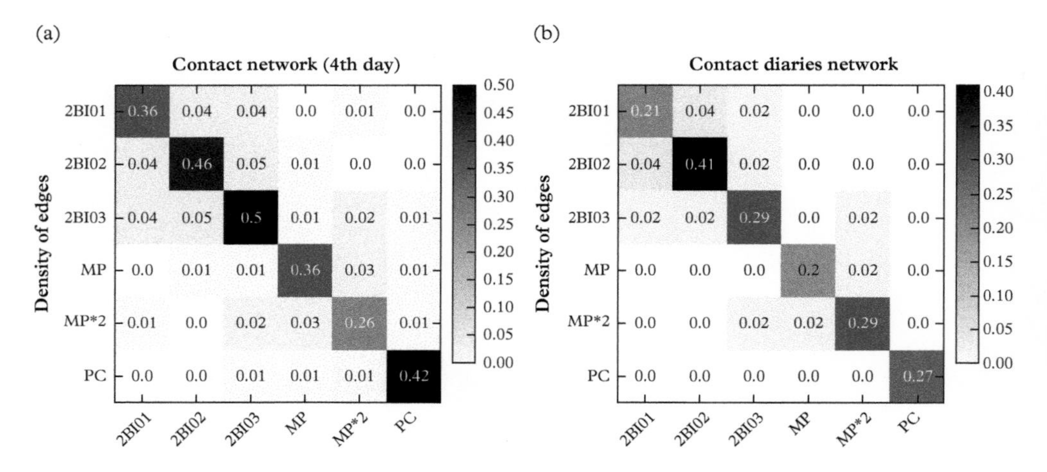


**Figure 6.4** *Contact matrices of link densities obtained from different data sources in a high school. We compare here the contact matrices of link densities between classes built from (a) the network of contacts obtained using the sensor data collected on a specific day and (b) the network of contacts as reported in the contact diaries collected on the same day. The similarity between these two matrices is 97%. From Ref. [189].*

but also by allowing a comparison of networks' structures at both global and local levels. The aggregated contact networks provide additional properties whose distributions can be measured and compared, such as nodes' degrees and strengths, and links' weights. The distributions of degrees (the number of distinct individuals with whom a given individual has been in contact) turn out to be similar across days and contexts, with narrow shapes, an exponential decay at large degrees, and characteristic average values that depend on the particular context [135]. The distributions of the cumulative contact durations are broad and very similar across very different contexts: different populations, in which individuals behave with very different goals in different spatial and social environments, display a strikingly similar statistical behavior. Finally, comparison of the neighborhoods of specific nodes on different days yields information about the rate of renewal of contacts between different days, an important quantity in the context of epidemic spreading phenomena. This rate turns out to be substantial but much smaller than if contacts were at random, and it has similar values across contexts.

Despite these statistical similarities, aggregated networks describing contacts in different contexts are obviously different, as revealed by a more detailed investigation. Differences in their structure can be already revealed by a visual inspection of simple force-based network layouts. For instance, the aggregated network of interactions during a conference day is much more "compact" than the ones describing the interactions between museum visitors. Aggregated networks of contacts among school children, high-school students, or office workers have a more modular structure [114, 189]. For instance, children in a single class form a cohesive structure with many links, but links between different classes, and in particular between children of different grades, are less frequent.

More subtle differences can be found by investigating the correlations between weights and network topology. In particular, we can consider for each node its degree (number of distinct individuals contacted) and its strength (cumulated time of interaction with other individuals). Correlations between these quantities are expected. For random durations of contacts a linear dependency of the average strength $\langle s(k) \rangle$ of nodes of degree $k$ is obtained; a superlinear dependence hints at the importance of superspreader nodes with large degree, while a sublinear behavior indicates that the decrease in the weights of individual contacts mitigates the expected superspreading behavior of large-degree nodes. In this respect, contrasting results have been obtained in different contexts, showing that coarse data summaries ignoring correlations might not carry enough information to fully characterize how diffusion processes would unfold on the contacts described by these data. Aggregated networks, even though they do not include detailed temporal information and remain a static description, shed, in this respect, important light on the system's dynamics.

### 6.3.2 Using detailed data in data-driven simulations

One of the most important practical uses of data consists in feeding data-driven models of dynamical processes such as, for instance, information diffusion or epidemic spread, which unfold on networks of communication or contacts. The issue of how much detail should be used when feeding such models is tightly linked with the above discussion on the advantages and limitations of data representations.

The answer depends in particular on the timescales of the process under investigation. For instance, when dealing with fast processes, the order in which events between nodes take place can be crucial in determining how fast the process spreads and how many nodes it impacts [111, 258]. On the other hand, for relatively slow processes such as simulations of realistic infectious disease spread, it was shown in Ref. [284] that aggregated networks can be used as a substitute for full temporal network data, under the condition that the aggregated network is weighted, that is, that the heterogeneous character of the interaction between individuals is taken into account. In the same spirit, Ref. [181] shows how the use of standard contact matrices in data-driven simulations of spreading processes can yield misleading results, while the CMDs, which provide a summary of the heterogeneity of contact patterns, can inform models of realistic infectious disease spread, by keeping the right amount of information and leaving out unimportant details, such as, for example, when one is not interested in who is specifically reached by the spread but rather in population level outcomes and in strategies based on grouping individuals according to their role or category in the population.

We discuss below two more practical uses of CMDs and of contact matrices in data-driven simulations.

#### *6.3.2.1 Using data representations to complement incomplete data*

As discussed above, comparison of various datasets describing contacts between individuals has revealed the strong robustness of the distributions of contact durations across contexts. Moreover, these distributions have been shown to be robust as well under sampling of the population under scrutiny. Contact matrices giving the density of links between groups of individuals are also unchanged by sampling. As discussed and shown in practice in Ref. [113], this can be a great resource when dealing with incomplete datasets.

Let us, for instance, assume that data describing the contacts between individuals has been collected in a given context, but that not all individuals have participated in the data collection. As a result, all information on the contacts involving the nonparticipating individuals seems to be lost. When simulating a spreading process on the data, these individuals effectively act as if they were immunized, as the potential transmission paths going through them cannot be taken into account. The outcome of the spread is therefore strongly underestimated [113] (see Figure 6.5).

The robustness of the CMD under population sampling means, however, that not all the relevant information about the nonparticipating individuals is unknown and that this data representation can be accurately measured, even when using the incomplete data. The CMD can then be used to create surrogate data, that is, sets of fake but realistic timelines of contacts among the nonparticipating individuals and between them and the participating ones. Using such surrogate data in the spreading simulations makes it possible to recover a correct evaluation of the epidemic risk [113], as shown in an example in Figure 6.5.

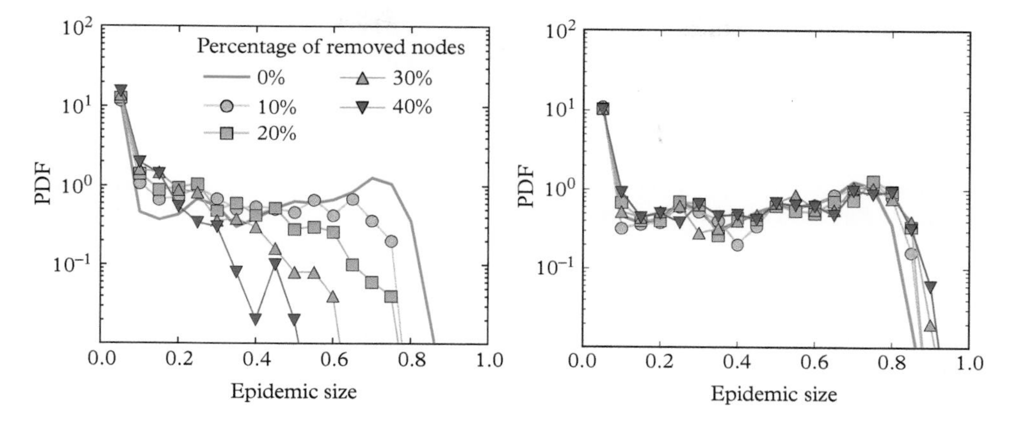


**Figure 6.5** *Distribution of epidemic sizes for a spreading process simulated on top of a temporal contact network—here, a high-school dataset. The red curve gives the distribution when the whole dataset is used. (Left) distributions obtained when data is missing. (Right) distributions obtained when surrogate data built using knowledge on the CMD measured on the sampled data is used. From Ref. [113].*

Another example of the usefulness of such representations has been given in Ref. [188]. Contact data obtained through contact diaries, due both to population sampling and to underreporting of contacts, is very incomplete and cannot be used easily in detailed simulations of spreading processes. However, the similarities of the contact matrices of link densities measured in both contact diaries and sensor data, together with the wide robustness of the distributions of contact durations measured in diverse settings, means that it is also possible to build, starting from the contact diaries, surrogate contact data that is similar enough to data from real contacts so that the outcome of the simulations of spread yields similar results [188]: once again, one uses the robustness of the CMD properties across data sources and contexts to measure the properties of the contacts most relevant for the issue at hand.

These two examples crucially confirm the usefulness of CMDs, by showing that the information they contain about the density of links between groups of individuals and the heterogeneity of contact durations is sufficient for use in simulations of realistic spreading processes.

#### *6.3.2.2 Mesoscale interventions*

Detailed, time-resolved networked data can yield a precise ranking of nodes according to their importance as measured, for instance, from their (temporal) betweenness centrality. Commonly, this type of ranking is used to identify nodes on which it would be interesting to act in order to mitigate or enhance a dynamical process (e.g., to mitigate an epidemic propagation). The ensuing individual-based control strategies, however, are difficult to carry out in practice. Moreover, in the case of temporally evolving networks, such rankings might be of limited efficiency, as the specific interactions among nodes do not repeat themselves in a precise way at different moments [283]. On the other side of the possible spectrum of data representations, too-coarse summaries of the data, such as global averages, offer only a very limited choice of control tools.

In this perspective, representations of the data at intermediate levels of detail can, in fact, carry crucial information suggesting possible efficient interventions at this intermediate scale. For instance, the contact matrix representation of the contact patterns occurring in a school shows that children spend much more time in contact with children of the same class and of their own grade. This is expected to be a rather general qualitative feature of schools, due both to age homophily and to schedule constraints, and suggests that transmission events might take place preferentially within the same class or grade. Hence, such data suggests targeted and reactive mitigation strategies in which one class or one grade is temporarily closed whenever symptomatic individuals are detected. As shown in Ref. [112], these strategies turn out to be almost as effective as whole-school closure, at a much lower cost in terms of service disruption.

Another example of mesoscale intervention is developed in Ref. [109] in the context of the decomposition of a temporal network of contacts in a sum of interpretable

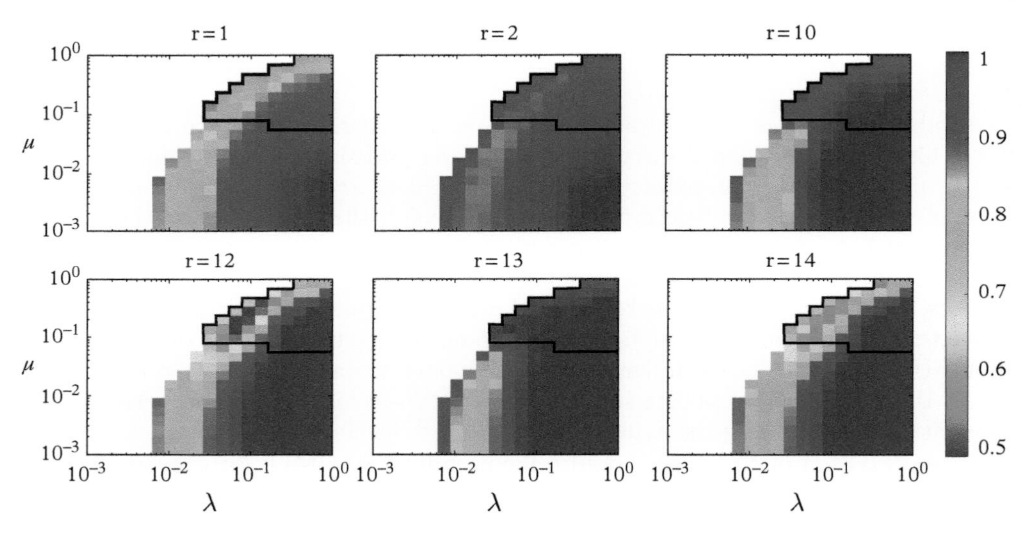


**Figure 6.6** *Impact of the removal of specific structures of a temporal network (school dataset) on an SIR spreading process. The heatmap shows the epidemic size ratio (ratio of epidemic sizes without and with intervention) as a function of the spreading model parameters. Each heat map corresponds to a targeted intervention that selectively removes one component. For each removed component* r*, the heat map shows the epidemic size ratio as a function of the SIR parameters* $\lambda$ *and* $\mu$*. Epidemic ratio = 1 indicates that the intervention does not affect epidemic size. The white area is the region where the epidemic dies out, that is, it fails to affect more than 1% of the network nodes. The region inside the black contour line corresponds to parameter values such that the SIR epidemic finishes within the finite span of the school dataset (2 days), both for the full and for the altered temporal networks. That is, for those parameter values, the epidemic size ratio is not affected by the finite temporal span of the dataset. The figure clearly shows the strong impact of removing components 12 or 14, which correspond to the mixing of classes during the breaks. From Ref. [109].*

components, thanks to the nonnegative tensor factorization [110] described in Section 6.2. Simulations of epidemic spreading processes with varying parameters can indeed be carried out either on the original temporal contact network or on a modified network in which a specific component $s$ has been removed (obtained by summing the other components), and the outcomes can be compared. When a component can be interpreted in terms of a specific behavioral pattern, its removal can be regarded as the effect of an intervention strategy that selectively targets that behavior. The case study presented in Ref. [109] corresponds to the interactions between children in a school. Its tensor decomposition yields, on the one hand, components corresponding to the classes and, on the other hand, components mixing different classes and corresponding mainly to the lunch breaks. As shown in Figure 6.6, the removal of the latter has a much stronger impact than the removal of the former, despite corresponding to a smaller number of links. Most importantly, such components could not be identified by traditional community detection methods but instead consist of weaker, temporally localized mixing patterns corresponding to scheduled social activities.

## 6.4 Conclusions and perspectives

The increase in data availability and resolution raises both opportunities and challenges related to their analysis, modeling, and practical use. Many datasets, in particular, are commonly used to feed data-driven models. To this aim, the right level of description needs to be found, one which keeps the relevant, salient properties of the data while discarding unnecessary details. Adequate data representations and null models need, therefore, to be defined. Naturally, different datasets can give rise to the same representation, once aggregated. It is thus interesting to define whole hierarchies of representations at intermediate aggregation levels. In this chapter, we reviewed some recent advances in the case of temporal networks and discussed their possible representations, from very detailed to very coarse. Each representation retains specific features of the data. For instance, higher-order temporal networks take into account non-Markovian aspects and preserve causality. CMDs keep information about the heterogeneities of contact durations between different groups. Mesoscale structures reveal complex interplays of activity and structural patterns. These representations can be used to compare datasets, which can, for instance, be similar at a certain level of aggregation but differ at a less aggregated one. They can also be useful to feed data-driven models of dynamical processes or generate synthetic datasets similar to a given, original one.

Many perspectives and issues remain obviously open. In particular, principled approaches to the design of hierarchies of data representations and null models are currently missing. Further techniques to detect structures and correlated activity patterns are needed. It is also important to devise minimal models at different levels of description that incorporate nontrivial longitudinal structures, mesoscopic structures, and correlated activity patterns. Finally, the issue of building approaches to dynamical processes on (temporal) networks by working directly at mesoscales remains an outstanding problem.

# 7

# Multilevel News Networks

**Borut Sluban[1], Jasmina Smailović[1], Miha Grčar[1], and Igor Mozetič[1]**

[1]Jozef Stefan Institute, Ljubljana, Slovenia

## 7.1 Introduction

### 7.1.1 Summary

This chapter describes how to construct time-varying, multilayer networks linking entities from online news articles. We demonstrate the approach on a collection of over 36 million news articles that were published around the world in the last 4 years. Our multifold approach identifies interesting events from thousands of daily news articles and models temporal interactions between the entities in the news. Informative news should answer the following questions: "*Who?*", "*Where?*", "*When?*", "*What?*", and, possibly, "*Why?*". The temporal aspect of the network answers the "*When?*" question, whereas the entity co-occurrence layer answers the "*Who?*" and the "*Where?*" questions. The summary layer answers the "*What?*" question, and the sentiment layer labels the links as "good" or "bad" news. We distinguish between usual/common and unusual/exceptional patterns in the news. We then compare the network to empirical, real-world networks and show that geographical proximity highly influences the co-occurrence of countries in the news and that countries with significant trade exchange tend to be jointly mentioned in a positive context. Finally, we propose an approach for identifying the most relevant events linking different entities and show that top news articles are not as positive as general news articles are. In addition, we have provided an interactive Web portal that demonstrates the evolution of the news network, the top news content, and the associated sentiment.

### 7.1.2 Motivation

News informs people about current events around the world. News articles mostly cover topics such as politics, business, sports, and extreme natural or social disasters but also

Sluban, B., Smailović, J., Grčar, M., and Mozetič, I., "Multilevel News Networks" in *Multiplex and Multilevel Networks*, edited by Battiston, S., Caldarelli, G., and Garas, A. © Oxford University Press 2019.
DOI: 10.1093/oso/9780198809456.003.0007

report on activities of various social groups or public personalities. The articles are spread and/or sold by various news agencies using different media. By monitoring news Web sites from around the globe, we analyzed both the structure and the content of news articles. While research in news analysis mainly addresses statistical properties and interlinking of news [179, 95, 169], we focus on the following research questions: (1) how can we extract both the usual, "everyday" patterns in the news and unusual, highly publicized events, (2) what do these types of news actually reflect, and (3) are there properties of the news that differ significantly between them?

We applied a set of text mining, sentiment analysis, and network analysis methods to answer these questions. The theory of complex networks characterizes systems in the form of entities (nodes) connected by some interactions (links) [5]. Since news articles talk about numerous different entities (persons, companies, countries, etc.) and their mutual interactions, they can be interpreted as a complex network. The methods of complex network analysis have strongly influenced and advanced research in social media, biology, and economics [54, 136]. In certain research areas, the available data does not have an inherent network structure like that for transportation networks, computer networks, or social networks. Depending on the available data and the field of research, various types of networks can be constructed, however. A special case of networks extracted from data are co-occurrence networks in which nodes represent entities (persons, companies, countries, etc.), and links represent an observation that these entities exist together in some data collection (e.g., a database or a news article) [84]. For textual sources, it is important to extract unambiguous entities through effective entity resolution [60] and to extract links between them, where the links represent real relations and are not created by chance. In our previous work, we developed a method to estimate the significance of co-occurrences, and a benchmark model against which their robustness could be evaluated [232]. This chapter builds on our preliminary research on the extraction of entity co-occurrence networks from news articles [271, 269] and extends it via a comparative analysis of usual and unusual events in the news, as follows. We constructed time-varying networks of entities appearing in news articles from around the world and then enriched the links between the entities according to textual context and sentiment, thus creating different network layers. By comparing the layers with different network comparison methods, we were able to draw interesting conclusions which answer some aspects of the research questions.

The difference between usual and unusual patterns in the news provides a baseline against which the properties of the news are compared. The essence of the usual news patterns is that, in such news, links between entities co-occur significantly more often than would be expected by chance. Consequently, the network of connected entities gradually varies through time. To see what shapes the usual, everyday news, we compared this network and the network of associated sentiment derived from it to three empirical networks constructed from real-world data, as illustrated in Figure 7.1.

In contrast, we identified unusual news as events causing significant deviations in the news volume over several weeks, for any pair of entities. The network of unusual events between entities, enriched by the most relevant news content and sentiment polarity, shown in Figure 7.2, was analyzed separately, and its properties were compared to those of the everyday-news network.

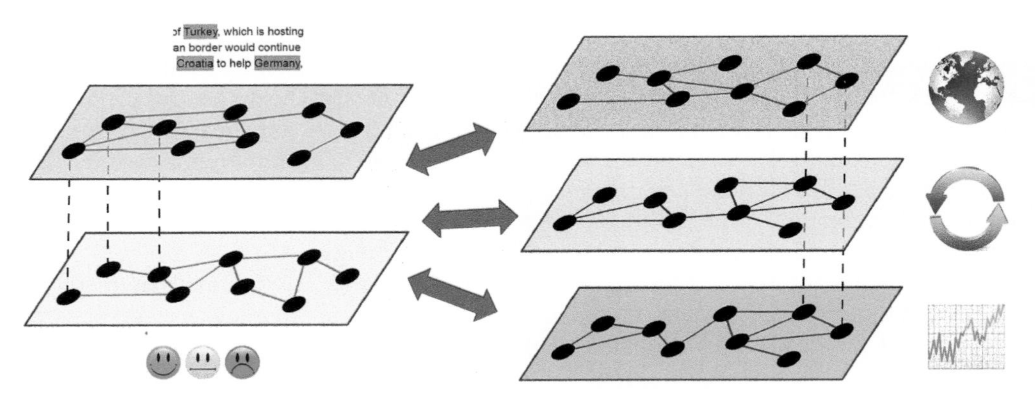


**Figure 7.1** *Analysis of everyday news. We compare two multiplex networks representing different types of relations between the same entities: significant co-occurrences and sentiment extracted from the news (left) versus geographical proximity, high trade, and high correlation of financial indicators (right, top to bottom). Adapted from Ref. [271].*

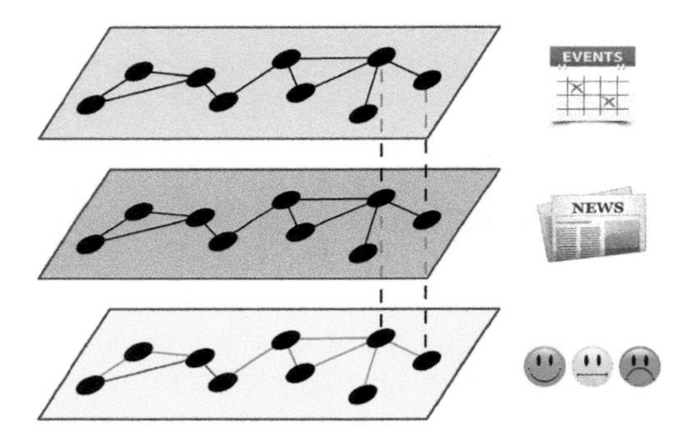


**Figure 7.2** *Analysis of unusual news—major news events between pairs of entities, accompanied by their context in terms of news content and the associated sentiment polarity.*

Our approach shows that geography and world trade influence the structure and sentiment of everyday news and that news in general tends to be slightly positive, whereas top news articles are more negative. Finally, we also discuss the presentation of such temporal multilayer news networks. The network evolution over time, with drill-down inspection of details, is demonstrated in a public, interactive Web portal at http://newsstream.ijs.si/occurrence/major-news-events-map. The portal facilitates access to over 38 million news articles, collected from 170 English news sites over the last years.

The chapter is organized as follows. Related work is presented in the next subsection. We describe the methods for modeling and analyzing news in Section 7.2. Section 7.3 presents the results on everyday news and major news events, their differences, and the implemented interactive visualization of news networks. We conclude in Section 7.4 with ideas for future work.

### 7.1.3 Related work

This chapter builds on the work of several research fields: complex network analysis, text mining, and sentiment analysis. In this subsection, we cover related research that is particularly relevant and provide references to broader overviews of the respective fields. The methods of complex network analysis are based on the methods developed in the fields of mathematics, computer science, and statistical physics, and they have strongly influenced and advanced research in social media, biology, and economics [54, 136]. Particularly interesting are co-occurrence networks, which are extracted from data in the form of entities of interest (nodes) and their relations (links), inferred from their common context. Co-occurrence networks are used in diverse fields, such as linguistics [84], bioinformatics [310, 62, 262], ecology [102], scientometry [184, 287], and socio-technological networks [57, 317, 116]. To extract networks from textual data, one needs to apply different text mining methods. First, detecting and disambiguating the entities of interest require efficient entity resolution [60]; second, the categorization of more complex relation types requires semantic analysis of the context [92]. Simple word co-occurrence networks have been used to model language structure [93] or to measure the relatedness between languages [178]. But textual data typically provides rich context to the entities it mentions, enabling the construction of various types of networks, such as signaling networks in biological systems, by extracting subject–predicate–object triplets [192], bisociative information networks for bridging concepts [142], or semantic networks for the purpose of text understanding and summarization [280, 193, 153, 264]. Sentiment analysis [177] can provide information on the emotional state of the entities, or attitudes expressed by other entities toward them [272, 238, 199]. Modeling the emotional dynamics in networks has been explored for interactions between users in social networks [194, 318], as well as for shared economic or political interests [273, 270].

Global news represents a rich resource of textual data, which has been used to extract different types of entity co-occurrence networks. Modeling personal connections evident from the news yields a social network of historically [220] or politically [296, 128] influential individuals and communities. Tracking country mentioning in global news was first used to identify the geographic community structure of the world's news media [172]. There exist several Web platforms which collect and analyze news articles, such as Lydia [179], NOAM [95], Event Registry [169], and European Media Monitor (EMM). The Lydia system provides various analyses for entities identified in news articles. For the selected entity, it constructs a relational network of relevant entities and provides different visualizations of popularity, sentiment, and geographical analysis (http://www.textmap.com/). The NOAM platform identifies topics, events, frequent phrases, and named entities in collected news articles. The system translates articles into English if they are written in some other language. Event Registry (http://www.eventregistry.org/) detects events and extracts main information about them from news articles written in various languages. The Web interface enables the search of events and exploration of corresponding articles, related events, and different visualizations. The EMM (http://emm.newsbrief.eu/) is a news aggregation and analysis system developed by the European Joint Research Centre. It provides identification of different named entities, topic aggregation over several languages, and article exploration.

## 7.2 Methods

We describe a multistage approach to the investigation of world news, combining text mining, network mining, and sentiment analysis. The stages are news acquisition and entity recognition; network construction; event detection; content identification; and sentiment analysis.

### 7.2.1 News network layers

Modeling the news requires monitoring entities of interest in the news, detecting their co-occurrences (links) over time, and identifying the associated context in terms of content and sentiment polarity. We present a method for the detection of significant co-occurrence links between entities and propose a time-aware method for detection of significant events about pairs of entities. These methods enable us to model the news as temporal networks connecting different entities. We propose also a method for identifying most relevant content of major news events and show how to assess the sentiment of a group of documents.

#### *7.2.1.1 News acquisition and entity recognition*

The news articles are collected by our data acquisition and processing pipeline implemented within the NewsStream platform (http://newsstream.ijs.si) [158]. The pipeline consists of several components for (i) data acquisition, (ii) data cleaning, (iii) natural-language preprocessing, and (iv) semantic annotation. The pipeline has been running continuously since October 2011, polling the Web and proprietary APIs for recent content and then turning it into a stream of preprocessed text documents. News is acquired from 2,600 RSS feeds from 170 English-language Web sites, covering the majority of Web news in English. On average, 25,000 news articles are collected per day. In the period from October 2011 to November 2015, more than 36 million unique documents were collected and processed.

News is about events related to individuals, social groups, countries, or companies, all of which we refer to as "entities." The process of identifying entities in textual documents requires three components: an ontology of entities and terms, gazetteers of the possible appearances of the entities in the text, and a semantic annotation procedure that finds and labels the entities. The ontology that we use for information extraction constitutes of three main categories: geographical entities, main protagonists (e.g., companies, politicians), and financial terms. Each entity in the ontology has associated with it a gazetteer, which is a set of rules that specify the lexicographic information about possible appearances of the entity in text. For example, "The United States of America" can appear in text as "USA," "US," "the United States," and so on. The rules include capitalization, lemmatization, POS tag constraints, must-contain constraints (i.e., another gazetteer must be detected in the document or in the sentence), and followed-by constraints. Finally, a semantic annotation procedure recognizes the entities of interest. It traverses each document and searches for entities from the ontology. The gazetteers of the entities in the ontology provide information required for the disambiguation of different appearances of the observed entities, resulting in a mostly correct annotation of the entities.

#### 7.2.1.2 *Significant co-occurrences: The co-occurrence layer*

Entities identified in a single piece of news (i.e., a document) can be connected with various types of relations. One of the simplest is their common appearance in the document, referred to as the co-occurrence of entities. Hence, for a selected set of entities $E = \{e_1, \ldots, e_l\}$, we construct a network layer of entity co-occurrences within a particular time frame—the *co-occurrence layer*. We use the significance algorithm proposed in Ref. [232] to assess whether the co-occurrence of two entities is statistically significant.

Let the number of all documents with at least two entities be $N$. Let $A$ and $B$ be two entities that occur with at least one other entity in $N_A$ and $N_B$ documents, respectively. Let $N_{AB}$ denote the number of the actual $A$ and $B$ co-occurrences. The expected number of co-occurrences is $\mathbb{E}(N_{AB}) = \frac{N_A N_B}{N}$. According to Ref. [232], the standard deviation is

$$\sigma_{AB} = \sqrt{\frac{N_A N_B}{N}\left(\frac{N^2 - N(N_A + N_B) + N_A N_B}{N(N-1)}\right)}, \tag{7.1}$$

and hence the standard significance score of the co-occurrence $N_{AB}$ from the data is

$$Z_{AB} = \frac{N_{AB} - \mathbb{E}(N_{AB})}{\sigma_{AB}}. \tag{7.2}$$

For a selected threshold $Z_0$, one can distinguish significant ($Z_{AB} > Z_0$) from non-significant ($Z_{AB} < Z_0$) co-occurrence relations between the two entities.

#### 7.2.1.3 *Major event detection: The event layer*

We use the daily volume of news documents as a proxy for identifying exceptional events in the news. Given a set of entities of interest $E = \{e_1, \ldots, e_l\}$, we identify all events related to all pairs of entities $(e_i, e_j)$. We monitor the volume of news about these pairs and construct a network of exceptional events between the observed entities—the *event layer*.

A link in the co-occurrence layer denotes that the number of actual co-occurrences is significantly greater than that expected by chance. The random co-occurrence baseline is estimated from the observed individual occurrences. Here we propose a different approach that compares the number of observed co-occurrences in a day to that from a longer time period.

We construct a time series of co-occurrence volumes $\mathbf{v}_{ij} = \{v_{ij}(t)\}_{t=0}^{T}$ for a pair $(e_i, e_j)$. At a given time point $t = p$, we consider a window $W_h(p) = \{v_{ij}(p-h-1), \ldots, v_{ij}(p-1)\}$ of length $h$ as a historical baseline, from which we calculate the expected volume at the time point $p$. We assume, for a pair of entities, that the volume of their co-occurrences in news is normally distributed around the average in a given time period. As the value of the average changes through time, we use the sliding window $W_h$ to adapt to recent changes.

Given the co-occurrence volume time series $\mathbf{v}_{ij}$, and the size $h$ of the historical data to be considered, we calculate the mean co-occurrence volume $\bar{v}_{ij}(p)$ in $W_h(p)$ and its standard deviation $\sigma_{ij}(p)$. Let $z_{ij}(p)$ denote the multiple of $\sigma_{ij}(p)$-deviations from the mean $\bar{v}_{ij}(p)$:

$$z_{ij}(p) = \frac{v_{ij}(p) - \bar{v}_{ij}(p)}{\sigma_{ij}(p)}. \tag{7.3}$$

The co-occurrence volume $v_{ij}(p)$ at the *peak* day $p$ is unexpected and represents an exceptional event occurring between the entities $e_i$ and $e_j$, when $z_{ij}(p) > Z_0$, for a given $Z_0$.

#### *7.2.1.4 Identification of the top news: The summary layer*

We attribute shallow semantics to the links in the network by a summary of the top news at peak days in the form of the most relevant headlines. First, we select all the news articles related to a particular link on a particular day. The headlines of these articles are then compiled into a single text document; one such document is created for each day in a 2-month period (excluding weekends). We then apply the standard text preprocessing approach to compute the bag-of-words (*BOW*) vectors of these documents [92]. In this process, we employ tokenization, stop word removal, stemming, and the *TF–IDF* weighting scheme [251]. The TF-IDF scheme is the most common weighting scheme used in text mining. The *TF* (term frequency) weight, $TF_{d,k}$, denotes the number of times the word $k$ occurs in the document $d$. The *IDF* (inverse document frequency) weight of the word $k$ is computed as $IDF_k = log\frac{|T|}{m_k}$, where $m_k$ is the number of documents in the collection $T$ that contain the word $k$. The *TF–IDF* scheme, $TFIDF_{d,k} = TF_{d,k} \times IDF_k$, weights a word higher if it occurs often in the same document (the *TF* component), and lower if it occurs in many documents from the corpus (the *IDF* component).

The *BOW* vector for the current day contains information about how important a certain word is with respect to the most relevant events on that day. Instead of showing the top-ranked words, we propagate the weights to the headlines and thus rank the headlines by their relevance. The weight-propagation formula is simple: we compute the average of the word weights in a headline $c$. The weight of the headline, $w_c$, is thus computed as

$$w_c = \frac{1}{|c|} \sum_{k \in c} TFIDF_{d^*,k}, \tag{7.4}$$

where $k$ enumerates the words in the headline $c$, and $d^*$ represents the merged documents for the day in question. Note that this technique tends to penalize long headlines. In our case, this is a desirable property because we would like to find short and to-the-point headlines that best describe the most important event(s). The most distinguished headlines at peak days represent the *summary layer* of the constructed network.

#### 7.2.1.5 Lexicon-based sentiment analysis: The sentiment layer

The sentiment polarity of a document is computed from the number of predefined sentiment terms (positive and negative) in the document. The sentiment terms are from the Harvard-IV-4 sentiment dictionary [291]. For a document $d$, the sentiment polarity $s_d$ is calculated as

$$s_d = \frac{pos_d - neg_d}{pos_d + neg_d}, \tag{7.5}$$

where *pos* and *neg* are the numbers of positive and negative dictionary terms found in the document $d$, respectively. The sentiment polarities of a set of documents can then be aggregated over time. The aggregate sentiment of a pair of entities $(e_i, e_j)$ in a certain time period $T$ is computed from the news documents $\{d, (e_i, e_j) \in d\}$ at days $t \in T$:

$$s_{ij}(T) = \frac{1}{n} \sum_{t \in T} \sum_{d \in t} s_d, \tag{7.6}$$

where $n$ is the total number of documents selected in the time period $T$. Based on the analysis of the sentiment distribution, we determine the thresholds $n_0$ and $p_0$ for the creation of positive and negative sentiment links, respectively.

### 7.2.2 Empirical network layers

We observe the same set of entities $E$ as in the "news network" layer, but the information regarding their mutual interactions is not acquired from the news. In particular, we explore three data sources to construct the empirical network layers: the geographical proximity of the entities, their interaction in terms of mutual trade, and correlations between their financial indicators.

#### 7.2.2.1 The Geo layer

The simplest among the "empirical network" layers is the geographical proximity layer, or "*Geo layer*". Each entity has a predominant geographical location, place of residence, address, or area. A link between two entities, $A$ and $B$, is established if a selected proximity measure is above a given threshold. Examples of proximity measures include geographical distance $d(A, B)$, inverse distance $\frac{1}{d(A,B)}$, or inverse squared distance $\frac{1}{d(A,B)^2}$.

#### 7.2.2.2 The Trade layer

The *Trade layer* models the interaction between entities as the amount of mutual trade. The amount of trade from $e_i$ to $e_j$ is denoted by $r(e_i, e_j)$. In total, $e_i$ is engaged in $r(e_i)$ worth of trade with all other entities, where $r(e_i) = \sum_{e_j \in E \setminus \{e_i\}} r(e_i, e_j)$. The cumulative, non-directed amount of trade between $e_i$ and $e_j$ is $r(e_i, e_j) + r(e_j, e_i)$. For $e_i$, its relative share of trade with $e_j$ is $R_{ij} = \frac{r(e_i,e_j)+r(e_j,e_i)}{r(e_i)}$. A trade link between two entities $e_i$ and $e_j$ is established if any of their relative trade shares, $R_{ij}$ or $R_{ji}$, is above a given threshold $R_0$.

#### 7.2.2.3 The Financial layer

Certain entities have an associated time-varying financial indicator (e.g., price, trade volume), which is represented as a time series $f$. A basic approach for measuring similar trends in the movement of financial indicators is the use of the Pearson correlation [226] between time series $f_i$ and $f_j$ of entities $e_i$ and $e_j$, over a period of $T$ time points:

$$\rho_{ij} = \frac{\sum_{t=1}^{T} (f_{i,t} - \bar{f}_i)(f_{j,t} - \bar{f}_j)}{\sqrt{\sum_{t=1}^{T} (f_{i,t} - \bar{f}_i)^2 \sum_{t=1}^{T} (f_{j,t} - \bar{f}_j)^2}}, \tag{7.7}$$

where $\bar{f}_i$ and $\bar{f}_j$ are the average values of the respective series. The *Financial layer* is constructed using a threshold value $\rho_0$, which determines whether the indicator time series of two entities are sufficiently correlated ($\rho_{ij} > \rho_0$) to form a link between them.

### 7.2.3 Network comparison measures

A comparison of network layers $\{L_1, \ldots, L_m\}$ can be done by measuring the link overlap between the layers. Let $l(L_a)$ and $l(L_b)$ be the sets of links in layers $L_a$ and $L_b$, where a link is defined as a pair of nodes it connects, for example, $(e_i, e_j)$. Then

$$o(L_a, L_b) = \frac{|l(L_a) \cap l(L_b)|}{|l(L_b)|} \tag{7.8}$$

is the size of their link overlap relative to layer $L_b$.

Considering for each layer not only the links but also their weights (strength of the relation), a comparison of the top strongest links in each layer can be adapted. Let $sl(L_a)$ and $sl(L_b)$ be sorted lists of links from layers $L_a$ and $L_b$, ordered by descending weights, and let $sl_k(L)$ denote the first $k$ elements of $sl(L)$. Then *precision at k* (*$prec_k$*) [237] is defined as

$$prec_k(L_a, L_b) = \frac{|sl_k(L_a) \cap sl_k(L_b)|}{k}. \tag{7.9}$$

If, for all pairs of layers $L_a$ and $L_b$, $a, b \in \{1, \ldots, m\}$, the same $k$ is selected, then a meta-network can be constructed with nodes representing layers $L_a$, $a \in \{1, \ldots, m\}$, and links representing the relation between the layers, where $prec_k(L_a, L_b)$ values are weights of the links, indicating the strength of the relationship.

Other comparisons of the network layers, induced on the "strongest" links for a particular relation type, are based on the most important nodes in each layer. In one approach, we measure the importance of nodes in terms of their *centrality*, as denoted by the *eigenvector centrality measure* [42]. Let $A$ be the adjacency matrix of nodes $e_1, \ldots, e_n$

in the network. The components of the eigenvector of the largest eigenvalue $\lambda$ solving the equation $A\mathbf{x} = \lambda\mathbf{x}$ hold the centrality values of the corresponding nodes. Nodes connected to better-connected nodes get higher centrality values. This measure can be used to compare the most central nodes between pairs of layers. Another approach to identify the most important nodes of a network is *k-core decomposition* [260]. This is an iterative process, pruning all the nodes with degree smaller than $k$. The remaining part of the network which holds only nodes with degree greater or equal to $k$ is called the $k$-core. The core with the largest $k$ is called the main core of the network. Comparing the main cores of different network layers can be used to assess the similarity between the layers.

## 7.3 Results and discussion

The proposed methods were used to (1) extract both usual, "everyday" patterns in the news and unusual, highly publicized events (2) analyze what usual and unusual news actually reflect, and (3) discover whether any properties of the news show significant differences between usual and unusual news. The results are presented in three parts. First, we analyze the everyday news; second, we focus on major news and how it differs from everyday news; and, finally, we show some visualizations of network snapshots.

### 7.3.1 Analysis of everyday news

#### *7.3.1.1 Network construction*

We monitored 50 countries in the news as entities of interest and constructed a network of their co-occurrences, using the significance algorithm presented in Section 7.2.1.2. From the news, we also created a sentiment layer of the country co-occurrences, showing the sentiment of the joint context in which both entities occur. The network layers were sampled in monthly snapshots over a time period of 2 years.

The construction of the empirical network, which should reflect the real-world context of the news, was done using data from external sources. For the Geo layer, we used the *is-a-neighbor-of* relation to link the selected countries. Links representing common terrestrial borders were extended also, with a few links between countries that are considered adjacent in the local geographical context, such as Australia and New Zealand, South Korea and Japan, and Italy and Malta.

Trading relations between countries were obtained from the UNCTAD website (http://unctadstat.unctad.org), the United Nations statistics data center, providing yearly aggregations of trade data. Our Trade layer was constructed from trade links that present relatively important trade relations (greater than 10%, i.e., $R_0 = 0.1$) for at least one of the connected countries.

We considered 50 countries that issue sovereign bonds and which are insured by credit default swaps (CDS), which is a type of insurance for when a bond issuer defaults and is unable to repay the debt. To construct the Financial layer, we used the time series of their CDS prices, which are often considered a good proxy for the risk of default of the country issuing bonds [223, 4]. We then created links between countries whose correlation between their CDS time series was $>0.9$ ($\rho_0 = 0.9$). In order to ensure that there was enough data for reliable correlation results, we used a 3-month time window for each snapshot, assigning it to the last month (e.g., the Nov–Dec–Jan window for the "Jan" snapshot).

#### *7.3.1.2 Co-occurrence versus empirical layers*

The results are presented for a multiplex network of 50 country nodes, for a 2-year time period: 2012 and 2013. The co-occurrence network, $L_{CO}$, varies with time, and we used a 1-month time window. The Geo layer, $L_{Geo}$, and the Trade layer, $L_{Tr}$, are static. The yearly aggregated trade data from 2012 was used for 2013 as well. The Financial layer, $L_{CDS}$, varies—we used a 3-month time window for it as well.

First, we present the analysis of overlapping links between the network layers $L_{CO}$, $L_{Geo}$, $L_{Tr}$, and $L_{CDS}$. We were interested in the number of links from the "empirical network" that appear in the news as country co-occurrences over time. The relative overlaps $o(L_{CO}, L)$ for $L \in \{L_{Geo}, L_{Tr}, L_{CDS}\}$ are presented in Figure 7.3. We see that most of the Geo layer links coincide with the country co-occurrences in the news, whereas, on average, less than half of the links between the countries in the Trade and Financial layers also appear in the co-occurrence layer.

Next, we investigated how sentiment is associated with the country co-occurrences related to the empirical network. Using the sentiment analysis approach presented in Section 7.2.1, we found that there is a strong bias toward positive sentiment in the news. We set thresholds $n_0$ and $p_0$ to two standard deviations away from the average sentiment polarity in the documents, thus selecting only links that reflect the most negative and most positive sentiment in the context of two countries. The negative sentiment layer turns out to be predominantly small, even for a slightly less restrictive threshold $n_0$ (at 90% standard deviation from the average) and therefore has mostly low overlap with the

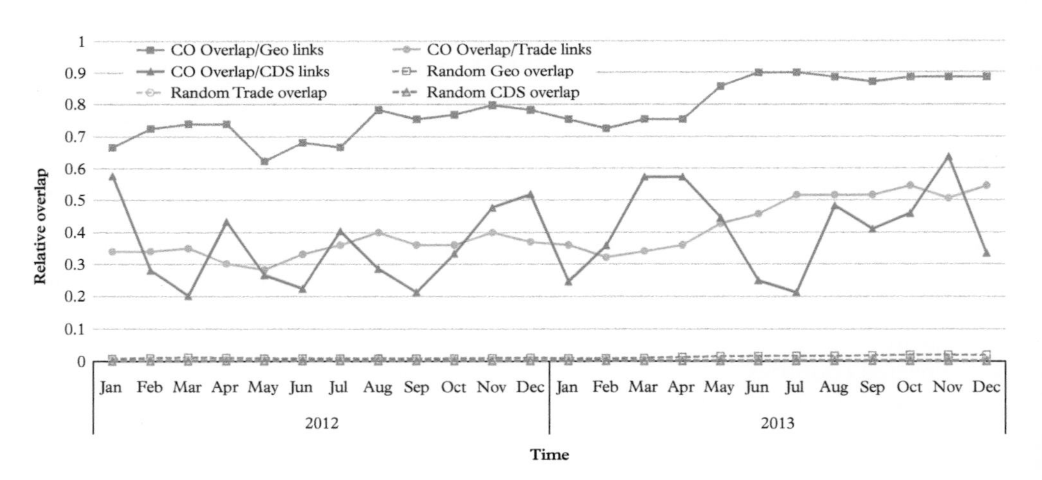


**Figure 7.3** *Relative amount of the empirical layer links present in the co-occurrence layer. From Ref. [271].*

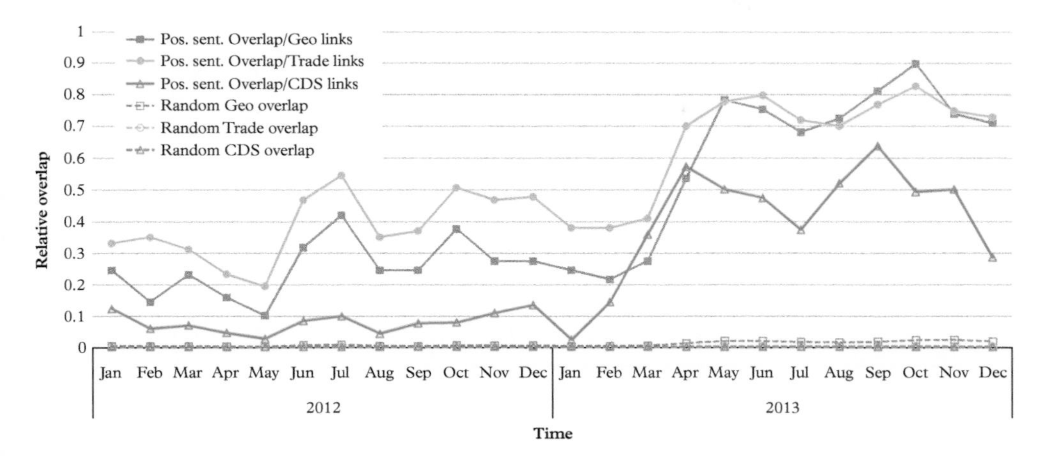


**Figure 7.4** *Relative amount of the empirical layers' links present in the positive sentiment layer. From Ref. [271].*

empirical layers. On the other hand, the comparison of the positive sentiment layer with the empirical layers results in a larger number of common links, as shown in Figure 7.4. Positive sentiment between the countries has the largest overlap with trade relations, followed by geographical proximity and, to the smallest extent, by the correlation between the CDS time series.

Comparison of the most important nodes in each layer shows similar results. Comparison of the main $k$-cores shows that the largest overlaps occur between the co-occurrence and the Geo layer cores, and between the positive sentiment and the Trade layer cores (see Figures 7.5 and 7.6). The co-occurrence layer cores overlap with the Geo layer cores in central European countries, and with the Trade layer cores in western European countries. The overlaps between the co-occurrence and the CDS layers show the common presence of some eastern European countries in 2012, but no regular presence in 2013. Several countries regularly appear in the core overlap between the positive sentiment and the Trade layers (CN, DE, US, UK, JP, BR, FR, and AU). Germany is also almost all the time (23 months) in the core overlap of the positive sentiment and the Geo layers.

Most central nodes of the co-occurrence layer coincide with the Geo layer in central European countries (AT, CZ, HU, SK, SI) and with the CDS layer in few eastern European countries; with the Trade layer, only Finland appears often among the top ten most central nodes. For the positive sentiment layer, the common most central nodes are Germany and Russia for the Geo layer, and some of the largest economies (CN, DE, FR, JP, RU, US) for the Trade layer.

Finally, we used the *precision-at-k* method to measure the link overlap of the strongest relations in each layer, in terms of the highest excess over random co-occurrence, most positive sentiment, highest mutual trade volume, and highest correlation between financial indicators. Limited by the number of links in the Geo layer, $k$ was set to 69.

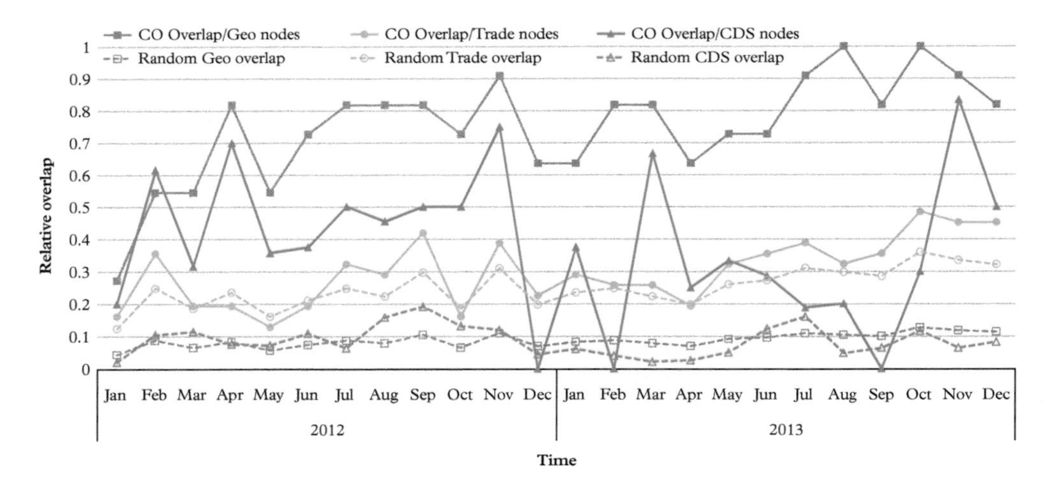


**Figure 7.5** *Relative amount of the empirical layers' nodes present in the co-occurrence layer when only the main k-cores are considered.*

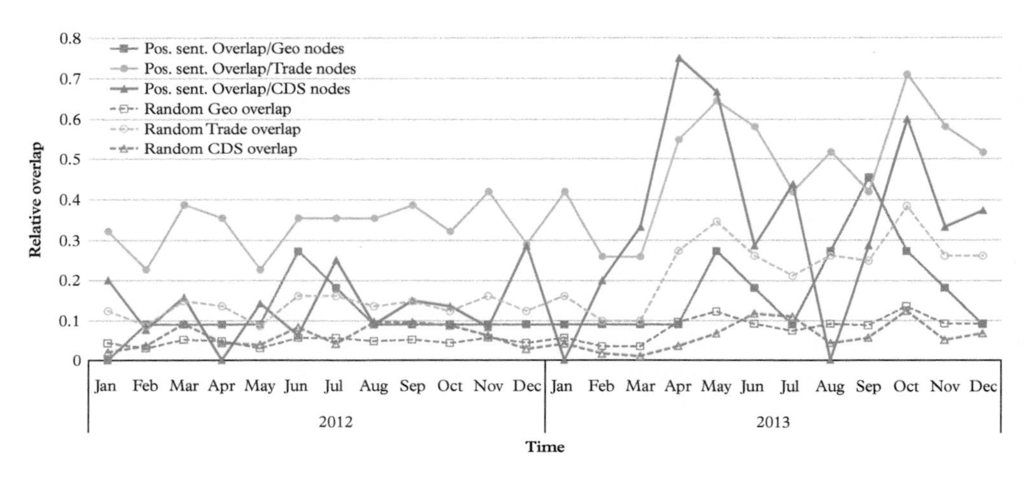


**Figure 7.6** *Relative amount of the empirical layers' nodes present in the positive sentiment layer when only the main k-cores are considered.*

Figure 7.7 illustrates the relations between the layers weighted by the *precision-at-69* values.

We see results that are similar to those in Figures 7.3 and 7.4. The News layer has the highest overlap of strongest links with the Geo layer. One can infer that neighboring countries tend to appear together in the news. On the other hand, the positive sentiment layer has the highest overlap with the Trade layer, suggesting that countries with high mutual trade tend to appear together in a positive context in the news. We can also observe a relatively strong relation of trade between neighboring countries, whereas relations between other layers are weaker.

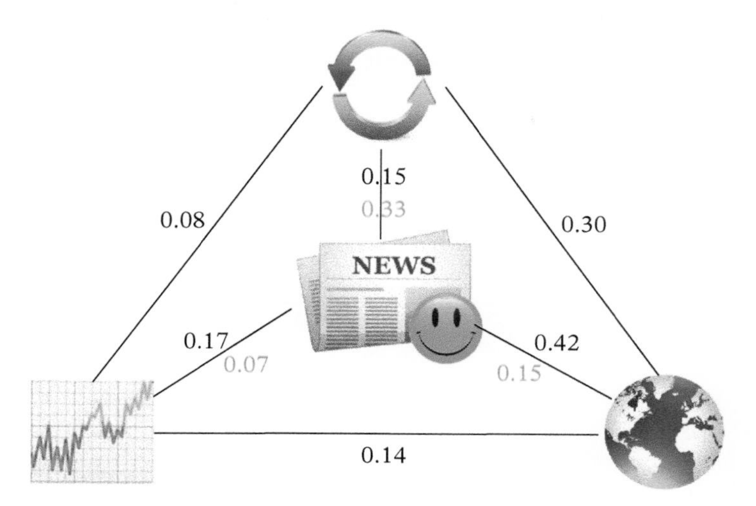


**Figure** 7.7 *A meta-network between the news (the central node) and the empirical network layers: the Trade layer (the top node), the Financial layer (the left node), and the Geo layer (the right node). The green numbers correspond to the positive sentiment layer. From Ref. [271].*

## 7.3.2 Major event news

In this section, we focus on the detection of major events in the news. We describe the construction of a temporal network of different countries as entities of interest, using news articles from the last 4 years. The network reveals the semantics of the relations between the countries in terms of the extracted contents and sentiment.

### *7.3.2.1 Significant events*

For all country pairs in the news on NewsStream, we selected news articles with at least three mentions of both entities and with at least one entity mentioned in the headline. We detected significant events by comparing the daily news volume to the volume of the news over the previous 2 months (44 weekdays; weekends were excluded due to their much lower volume of news). We assumed a normal distribution of the entity co-occurrence volume around the average number of co-occurrences over the 2-month period. We set $Z_0 = 3$ to identify most outstanding news production increases about a pair of entities. Hence, if on a particular day the news volume exceeded the average volume of the past 2 months by more than 3 standard deviations, this day was identified as a significant event day—the peak day for the observed pair of entities.

In Figure 7.8, we show the volume of news containing the entities "China" and "United Stateds" in the period between November 2011 and October 2015. The significant increases in the news volume are peaks above the gray line, indicating $\bar{v}_{\text{CN–US}} + 3 \cdot \sigma_{\text{CN–US}}$.

We added labels to some of the volume peaks in Figure 7.8 to illustrate what kind of significant news events related to China and the United States happened on the

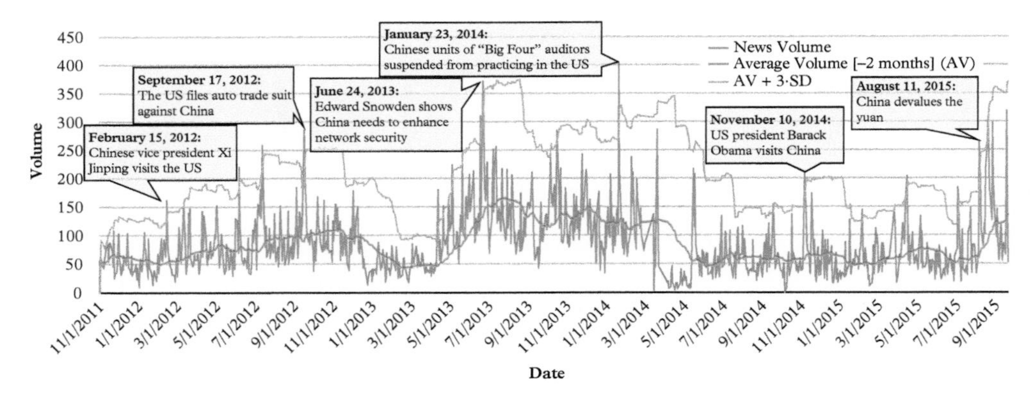


**Figure 7.8** *Volume of news articles about "China" and "United States." Significant events are at days peaking more than 3 standard deviations above the average volume in the previous 2 months. Labels at certain peak days describe an event concerning China and the United States on that day, as manually identified in the news. Adapted from Ref. [269].*

particular dates. These facts were obtained by manually inspecting news and official press releases of both countries. In the period between January 2012 and October 2015, 17,997 significant events between 217 countries were detected. We analyze these events in terms of the automatically detected relevant content and the associated sentiment in the following subsections.

#### *7.3.2.2 Contents and sentiment*

We identified the most relevant and distinguishing topics for each significant event day, as described in Section 7.2.1.4. However, evaluating the obtained results in terms of their relevance proved to be quite challenging, as there was no exact ground truth for the type of events that we were detecting. One publicly available resource of historical major news events that we could find is provided by the EMM.

Hence, we compared our top news results to the major news timeline of EMM, http://emm.newsexplorer.eu/NewsExplorer/timelineedition/en/timeline.html. The overlap with all the EMM major news was 45%, and 60% with major news topics mentioning at least two countries in the topic title. These differences are mostly due to the following reasons. As major news events of EMM are not limited to country relations (links), they include also news events mentioning only one country or none at all. Our approach, on the other hand, detects a wider range of more specialized events, which relate pairs of countries (entities). Topics persisting for several days with low evolution are avoided by our approach, as we are looking for significant new events. A country's involvement in a certain topic may be overlooked in the evaluation process due to unresolved indirect mentioning, such as the use of the terms "Merkel" or "VW" instead of Germany.

Some significant event days in August and September 2015, for four country pairs, are presented in Tables 7.1 and 7.2. Each event is characterized by the top news headlines detected by our approach, and the associated sentiment calculated from the texts of the top news.

The first example describes the events related to significant increases of news volume about China and the United States in August 2015. As China devalued its currency, the *renminbi* (or *Chinese yuan*, as it is better known internationally), the US media discussed the possible impacts on its economy. Similar increased media coverage of the two countries can be observed when the effects of China's actions become evident and when China cuts its interest rates. Notice the changes in the sentiment of these news, ranging from neutral (initially) and negative (perceived effects) to less negative (relief after shock). The second two examples are about events concerning French-built warships which were intended for Russia, but were later sold to Egypt. These events are also accompanied by different sentiment polarity. The fourth example highlights the "emissions scandal" of the German automobile producer Volkswagen (VW) as it broke in the United States in September 2015.

We constructed the *major event network* from significant news events occurring between country pairs. We used the top three most relevant news articles at those peak days and their associated sentiment to construct the summary and sentiment layers.

**Table 7.1** *Content and sentiment of the most relevant news articles on significant event days involving China and the United States, in August 2015.*

| Link | Day | News Headline | Sentiment |
|---|---|---|---|
| **CN–US** (−0.189) | Aug 11 2015 | China Devalues Yuan | −0.022 |
| | | China Devalues Renminbi | 0.139 |
| | | China Devalues Yuan by 2% | 0.056 |
| | Aug 12 2015 | China Devalues Its Currency Again | −0.333 |
| | | China Currency Falls for 2nd Day after Surprise Devaluation | −0.333 |
| | | China Currency Falls Again for 2nd day after Surprise Devaluation | −0.228 |
| | Aug 24 2015 | Alarm Bells Ring as China Sinks, Dollar Tumbles | −0.290 |
| | | Great Fall of China Sinks World Markets | −0.356 |
| | | Great Fall of China Sinks World Stocks, Dollar Tumbles | −0.300 |
| | Aug 25 2015 | Global Markets Rebound after China Cuts Rates | −0.193 |
| | | Global Markets Rebound after China Cuts Interest Rates | −0.234 |
| | | Dow Jumps 300 Points as China Cuts Interest Rates | −0.171 |

**Table 7.2** *Content and sentiment of the most relevant news articles on significant event days involving France and Russia, France and Egypt, and Germany and the United States, in August and September 2015.*

| Link | Day | News Headline | Sentiment |
|---|---|---|---|
| **FR–RU** (0.265) | Aug 6 2015 | France To Pay Russia under $1.31 Billion over Warships | 0.286 |
| | | France To Pay Russia under 1.2 Billion Euros over Warships | 0.256 |
| | | France Says Several Nations Interested in Mistral Warships | 0.254 |
| **FR–EG** (−0.072) | Sep 23 2015 | France Sells 2 Disputed Warships to Egypt | −0.091 |
| | | France Sells Warships to Egypt after Russia Deal Scrapped | −0.020 |
| | | France To Sell Warships to Egypt after Russia Deal Scrapped | −0.103 |
| **DE–US** (−0.015) | Sep 21 2015 | VW Rocked by US Emissions Scandal as Stock Slides 17 Percent | 0.039 |
| | | VW Rocked by U.S. Emissions Scandal as Stock Slides 17% | −0.036 |
| | | VW Shares Plunge on Emissions Scandal; US Widens Probe | −0.026 |
| | Sep 24 2015 | Will Volkswagen Scandal Tarnish Made in Germany Image? | 0.007 |
| | | After Year of Stonewalling, Volkswagen Stunned U.S. Regulators with Confession | −0.042 |
| | | Insight: After Year of Stonewalling, Volkswagen Stunned U.S. Regulators with... | −0.030 |

#### *7.3.2.3 Comparison to everyday news*

We compared the "everyday" news and the "major event" news in terms of network structure and the sentiment of the news. For the comparison between the everyday and major event news networks, the links of the event network were merged for each month and filtered to the selected 50 countries. Details of the structural comparison are presented in Table 7.3. Structural analysis shows that the major event network is, on average, less densely connected, having less than 40% of the links in the everyday network. On average, only a quarter of the event links are also in the everyday network; this is due to the greater sensitivity (daily resolution) of the event detection approach, which can identify individual peak days that are left undetected by the averaging over

**Table 7.3** *Comparison of the structural properties of the everyday news and the major event news networks. Shown are the average values with standard deviations of network density, clustering coefficient, and network assortativity, over a period of 2 years.*

| Network | Density | Clustering | Assortativity |
|---|---|---|---|
| Everyday news | $0.23 \pm 0.07$ | $0.65 \pm 0.06$ | $0.53 \pm 0.07$ |
| Major event news | $0.12 \pm 0.03$ | $0.27 \pm 0.12$ | $-0.22 \pm 0.10$ |

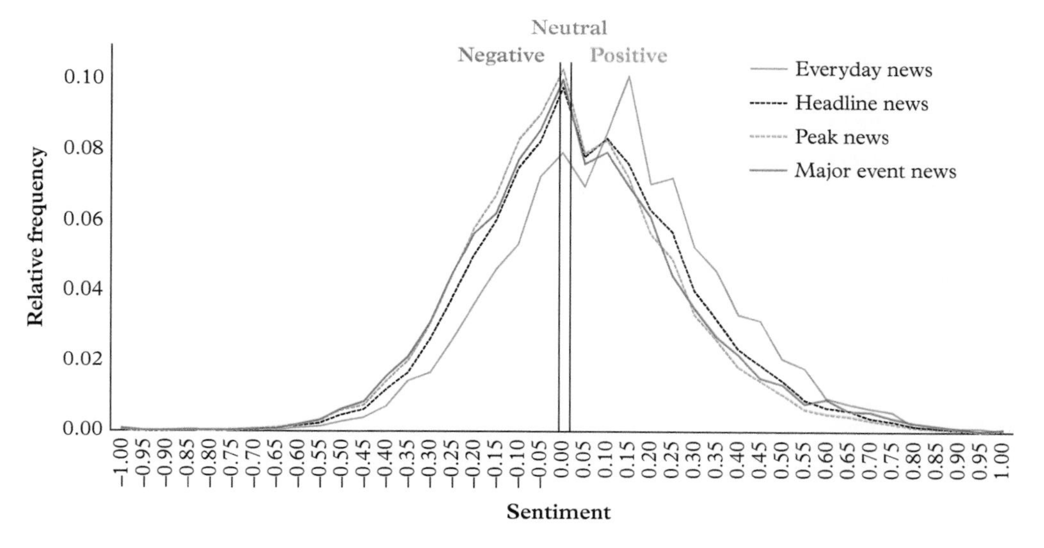


**Figure 7.9** *Comparison of sentiment distributions of everyday news articles, headline news, peak day articles, and major event news, that is, the most relevant articles on peak days.*

several days, as was done in Ref. [232]. The major event network also has a lower clustering coefficient, which suggests that major events in the news tend to involve fewer countries than everyday news does. The most interesting structural difference between the two networks concerns their assortativity [208]. In the everyday network, countries with similar presence in the news tend to co-occur, whereas, in the major event network, unequally represented countries tend to co-occur, which supports the unusual nature of the detected events.

We next examined the differences in the sentiment distribution of everyday news and the major event news. We also included the "headline" news, articles which are basis for event detection and which mention at least one of the relevant countries in their headline, and the "peak" news, that is, all the news on significant event days. Figure 7.9 shows the sentiment distributions.

All four sentiment distributions are approximately normal, and very similar. There is an evident positive sentiment bias in everyday news, while peak news is slightly negative. Headline news also shows a minor positive bias, whereas major event news is, on average, less positive but contain proportionally more extremely positive and extremely negative news articles. These statistics are summarized in Table 7.4.

We tested the null hypothesis that a pair of news populations has equal mean sentiment. We applied Welch's $t$-test [309], which is robust for skewed distributions and large sample sizes [91]. The results are shown in Table 7.5. With $t$-values $>10$, degrees of freedom $\gg 100$, and the $p$-value $\approx 0$, the null hypothesis can be rejected for all pairs of news populations. We conclude, with high confidence, that the four populations of news have significantly different sentiment means, although some of these differences are very small.

We then introduced a neutral zone around the sentiment mean $\bar{s}$, to distinguish "bad" news from "good" news. As $\bar{s}$ is the sample mean, the population mean lay in the interval $\bar{s} \pm 9SEM$, with very high confidence. We classified the sentiment of the top news into three discrete classes: *negative* if $-1 \leq s < 0$, *neutral* if $0 \leq s \leq 0.02$, and *positive* if

**Table 7.4** *The sentiment distributions for different sets of news relating pairs of countries. "Documents" is the number of documents, $\bar{s}$ is the sentiment mean,* SD *is the standard deviation, and* SEM *is the standard error of the mean.*

| News corpus | Documents | $\bar{s}$ | *SD* | *SEM* |
|---|---|---|---|---|
| Everyday news | 7,391,204 | 0.092 | 0.243 | 0.0001 |
| Headline news | 1,590,388 | 0.029 | 0.239 | 0.0002 |
| Peak news | 279,432 | −0.002 | 0.232 | 0.0004 |
| Major event news | 48,864 | 0.011 | 0.249 | 0.0011 |

**Table 7.5** *The results of Welch's* t*-tests for comparison of sentiment means;* DF *stands for the estimated degrees of freedom.*

| News Corpora | $t$ | *DF* |
|---|---|---|
| Everyday news vs. headline news | 300.56 | 2,355,313 |
| Everyday news vs. peak news | 209.88 | 303,211 |
| Everyday news vs. major event news | 71.64 | 49,480 |
| Headline news vs. peak news | 64.78 | 391,099 |
| Headline news vs. major event news | 15.83 | 51,655 |
| Major event news vs. peak news | 10.60 | 64,477 |

$0.02 < s \leq 1$. The neutral zone was used to distinguish between negative and positive sentiment for major event news in the network visualization.

#### *7.3.2.4 Network visualization*

Network visualization offers a way to better understand and analyze complex systems by enabling the user to easily inspect and comprehend relations between individual units and their properties [241]. As well as single-layer network visualization [25, 23], multilayer network visualization is becoming increasingly popular for highlighting various aspect of complex systems [259, 161, 71, 231].

We implemented a spatiotemporal visualization of the country co-occurrence network, constructed from major event news, their most relevant content, and the associated sentiment. The visualization was implemented within the NewsStream portal, to facilitate the inspection of various aspects of the network: time dimension, news content, news sentiment, and geography. The network was then embedded into the world map; the interface includes functionalities for exploring different aspects of the network. Figures 7.10 and 7.11 show two instances of the network in time and space. The visualization is an extension of the NewsStream portal and is publicly accessible at http://newsstream.ijs.si/occurrence/major-news-events-map.

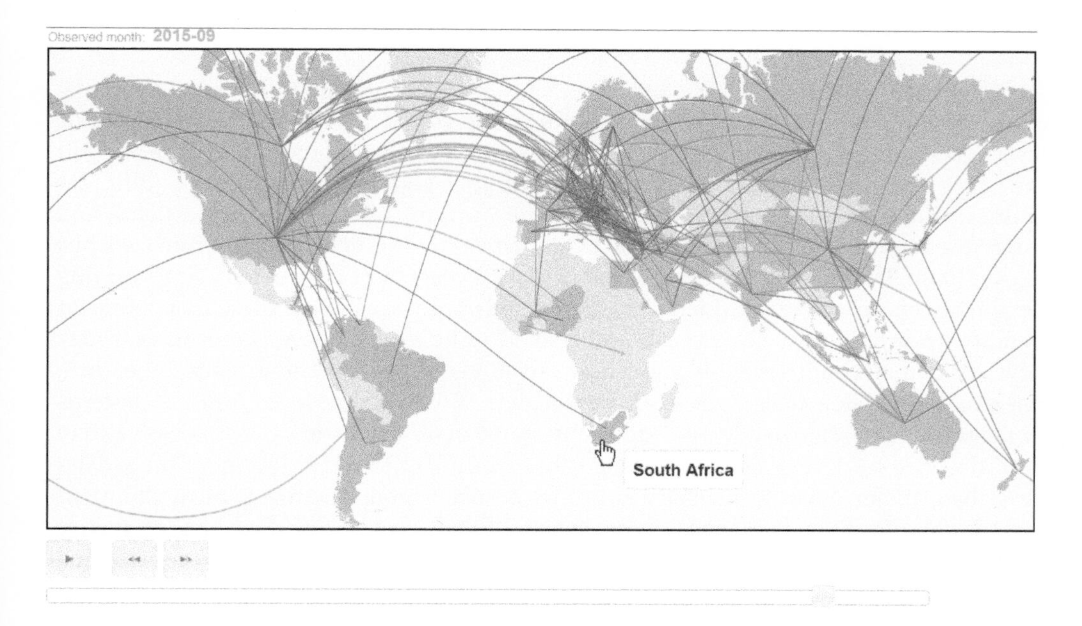


**Figure 7.10** *Temporal country co-occurrence network of major news events during September 2015. From Ref. [269].*

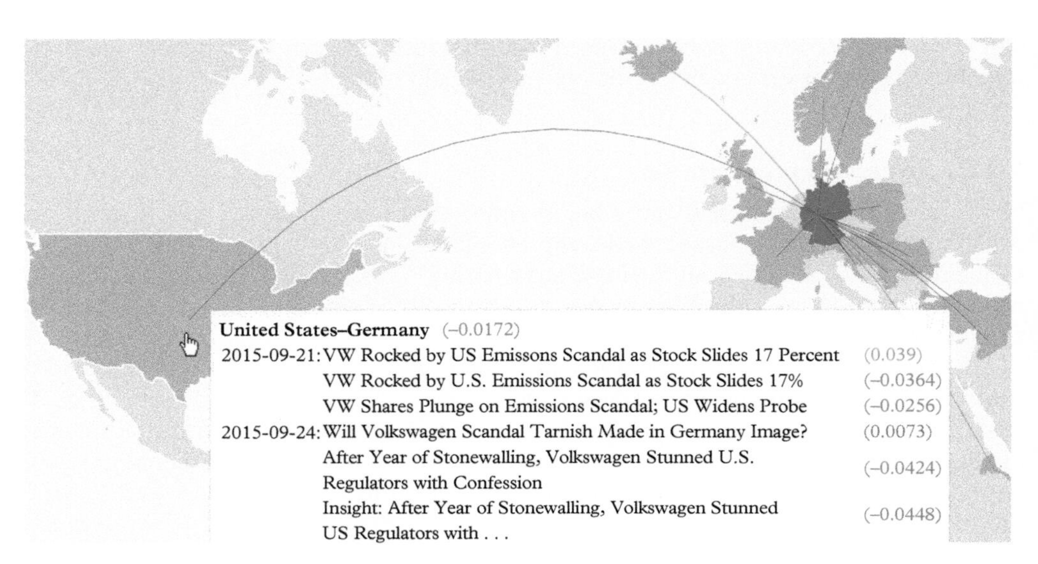


**Figure 7.11** *The most significant news articles involving Germany and the United States in September 2015. From Ref. [269].*

## 7.4 Conclusions

In this chapter, we described several methods for modeling and analyzing news by means of text mining, network mining, and sentiment analysis. The resulting temporal multilayer news network revealed the dynamic relations between various entities appearing in the news. It made it possible to capture both everyday and unusual news about different entities, summarize the relations between them in terms of most relevant articles, and assign the sentiment of the various articles to their corresponding context. From a corpus of over 36 million news articles published in the last years, we constructed a time-varying network between countries which were mentioned in the news. We showed that countries that are geographically close to each other tend to be mentioned together in everyday news, and that countries with good trade relations tend to mention each other in a positive context in articles from their newspapers. When we explored unusual patterns in news, on the other hand, we found that major news events are more negative than everyday news, which has an evident positive sentiment bias. Furthermore, in articles describing major news event, there are more co-occurrences between unequally mentioned countries than in everyday news articles. Finally, we implemented an interactive network visualization that supports the spatiotemporal exploration of the constructed networks.

We plan to broaden the range of semantic relations extracted from text in order to construct a public knowledge network from news. Another direction of future research is studying the role of news in the policy-making process. As news shapes opinion in policy debates, we plan to extend the news network with (in)direct ownership structure of the media companies and analyze how this influences the reported news.

## ACKNOWLEDGMENTS

This work was supported in part by the European Commission FP7 project MULTIPLEX (no. 317532), and by the Slovenian Research Agency (research core funding no. P2-103).

# 8

# The Role of Local Interactions in Cities' Global Networking of Multinational Firms: An SIR Model Applied to Partial-Multiplex Directed Networks

**Maria Tsouchnika[1], Michael Kanetidis[1], Celine Rozenblat[2], and Panos Argyrakis[1]**

[1]Department of Physics, University of Thessaloniki, Thessaloniki, Greece
[2]Faculty of Geosciences and Environment, University of Lausanne, Lausanne, Switzerland

## 8.1 Introduction

Global firms interact with each other to form complex networks of financial relations of ownership relations between them [305]. Whatever their activities, the networks of companies are mostly concentrated in the main global cities of the world, where they benefit from human, natural, and financial resources, but, reversely, firms' networks contribute to build the global characters of cities [255]. Such global cities are often the place where firms exchange many local financial linkages: the major example in the world is London, where more than 70% of the firms' linkages are between firms located inside London's urban region [244]. The average for all cities of the world shows that more than 35% of all firms' linkages are local. These local financial linkages reveal crucial issues for firms and for cities hosting these firms:

- If firms are in the same activity sector, local financial linkages often reveal local alliances between firms forming specialized economic clusters in cities: the financial

Tsouchnika, M., Kanetidis, M., Rozenblat, C., and Argyrakis, P., "The Role of Local Interactions in Cities' Global Networking of Multinational Firms: An SIR Model Applied to Partial-Multiplex Directed Networks" in *Multiplex and Multilevel Networks*, edited by Battiston, S., Caldarelli, G., and Garas, A. © Oxford University Press 2019.
DOI: 10.1093/oso/9780198809456.003.0008

linkages consolidate subcontracting long-lasting relationships or collaborations in innovation between firms.

- If firms are in different activity sectors, this could indicate different economic strategies: either the firms diversify their assets or the firms benefit from local capital investments (from banks or insurance).

In both cases, those local linkages constitute both the strength and the weakness of cities: if some firms fail, then the others have a high likelihood of failing. This is what happened in London in 2008 during a period of a few months after the big financial crash, when a cascade of failures resulted in thousands of employees losing their jobs. One may ask, to what extent do local interactions of firms of similar or different activities inside cities play a role for the diffusion of crises between cities?

To explore these questions, we simulate the spreading of a financial crisis in a partial-multiplex, directed network, noting that it is crucial to distinguish between the effects of two important factors that shape the relationships between the cities; their geographical proximity (effects of gravitation model on intercity flows of ownership links; see [245, 305]) and their activity proximity (exchanging many ownership linkages in close activities). Inside cities, the relations of different activities reflect the "related variety" of each city. By applying an SIR process to this network, we examine the possible outcome of the spreading of a catastrophic event. Specifically, we explore how both the speed and the peak of the propagation procedure vary with the layer from which the spreading started. We investigate possible differences between taking into account or not co-location of firms with different activities in the same cities. This then provides about the role of intracity "related variety" effects on the whole intercities network [103, 64]. The consideration of the intra-urban (mesolevel) role on the inter-urban (macrolevel) dynamic would be a first result of a multilevel effect in a multilayer network.

## 8.2 Data

We constructed a weighted and directed network of 800,000 companies (network nodes) forming multinational firms, with respect to their 1.2 million ownership relations (network links) (from the 2013 UNIL–GeoDivercity–Orbis database) and extracted the corresponding directed network of the cities that harbor these firms. In order to base the evaluation of cities' interactions on strong and stable networks, we use firms' ownership networks (internal networks) representing "observed" financial links between companies and creating "quasi-trees" [6, 306]. We built a large database of all direct and indirect linkages of financial ownership developed directly or indirectly by the top 3,000 worldwide companies according to their turnover in 2009 (for linkages observed in 2010) and in 2012 (for 2013). These data are derived from the best possible source of information: the ORBIS database (Bureau van Dijk, 2010, 2013).[1] The University

[1] This database encompasses about 700,000 subsidiaries from 2010, and 800,000 subsidiaries from 2013, with 1 million ownership links in 2010, and 1.2 million in 2013. The first 3,000 firms comprise different sets

of Lausanne and the European Research Council Advanced Grant GeoDiverCity completed the data by focusing especially on the locations and activities of firms. The two teams developed an original method to measure the geographical aggregation of firms within "large urban regions" (LUR) at the world scale. The spatial units chosen are the extended functional urban areas that are defined for each country in the accessibility zone around the major international airports [246]. This makes it possible to compare cities globally within delineations having the same urban significance all over the world. We will refer to them in this chapter as LURs or, for the sake of simplicity, "cities."[2]

A firm's network is a graph that consists of headquarters and subsidiaries as nodes that are linked with the oriented financial linkages (according to ownership) observed between them. The strength of these linkages is always above 10% of the capital of the subsidiary being owned by the headquarters (or another subsidiary from that firm). In order to shift from these individual firms' networks toward cities' networks, and as the exact proportion of owned capital is not always given with enough precision, we measured the intensity of a city relationship with any other city by adding the total number of financial relations of the firms located in the pair of LURs: $LO_{ij}$ is the number of subsidiaries located in the city $j$ having their headquarters (or a minority shareholder) in the city $i$. All the oriented linkages $l^e_{ij}$ of the enterprises $e$ having a shareholder or a headquarters in city $i$ and a subsidiary located in city $j$ count for 1 and are summed up to obtain $LO_{ij}$:

$$LO_{ij} = \sum_e l^e_{ij}, \tag{8.1}$$

where the summation is over all enterprises $e$, having headquarters in city $i$ and a subsidiary in city $j$. Note that the city $j$ can also have enterprises $f$ hosting shareholders who invest in subsidiaries of the city $i$. We use these oriented linkages in the following study we will use these oriented linkages, distinguishing the collection of inter-urban linkages, $LO_{ij}$, from that of intra-urban linkages, $LO_{ii}$.

The firms' networks are then classified into five categories according to their main field of activity, that is, high tech, low tech, knowledge-intensive services, less knowledge-intensive services, and others [196]. We employed this classification to divide the network into activity-specific relation networks that are connected through the cities they have in common, thus constructing a weighted, directed, multilayered, partial-multiplex network of cities. Then, we built two sets of multilayered networks, with each layer representing the network of cities for one activity sector; thus, cities are linked to each other not only within each layer but also between layers (see Figure 8.1).

in 2010 and in 2013, with a stable common core but different networks subsidiaries. These two independent sets of data make it possible to evaluate the power of integration of the global networks at each date, taking into account the transformations of the dominant economic actors.

[2] This approach requires significant effort and a high degree of expertise in each country in order to accurately evaluate the relevance of the LURs' delineations.

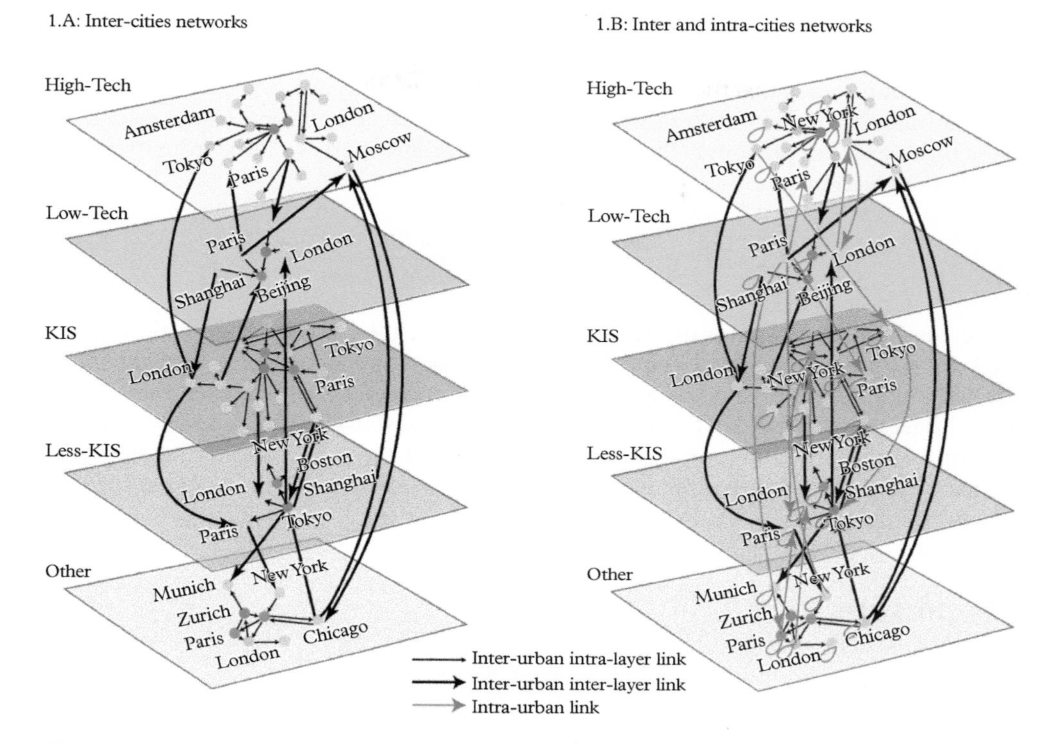


**Figure 8.1** *The multilayer city networks of multinational firms, by activity sector.*

In one case (Figure 8.1.A), we only take into account the linkages between cities (representing 36,952 pairs of cities and 389,749 firms' linkages), while, in the second case (Figure 8.1.B), we add the intra-urban ones (intra- and interlayers): 37,362 pairs of cities and 632,550 firms' linkages. Despite the fact that there are only 410 intra-urban reflexive linkages (1% of the pairs of cities), these reflexive linkages represent 242,801 firms' linkages (38% of the total of firms' linkages).

## 8.3 The simulation method

We utilize standard models of spreading phenomena in order to simulate the spreading of a crisis. Briefly, we start at a node that is chosen randomly and infect it at time $t = 0$. Then, the crisis propagates in one time step to all its nearest neighbors, using the well-known breadth-first search algorithm [268]. Then, the process is continued in time, the crisis propagates to new neighbors, and so on. Nodes that are designated as susceptible (S) are healthy but can be infected. Nodes that are infected (I) can spread the crisis to their neighbors. Nodes that are recovered/removed (R) are ones that had previously been infected but are not currently infected and cannot be infected again. Variations of the model include SIR, SIS, SIRS, and so on. In the present work, we use the SIR model [146]. As we have 5 layers, if one is to include all possible interactions, then there are 25 sets of interactions, the nodes of each layer interacting with the ones on the same layer (intralayer interactions), as well as with the nodes of each one of the four other layers (interlayer interactions). The probability ($q$) for infection of a node from its nearest neighbor is given by

$$q = k\frac{\textit{weight of the directed link}}{\textit{total weighted degree of target node}}, \tag{8.2}$$

where $k$ is an intensity factor in the range $0 < k < 1$ for the interlayer interactions, while $k = 1$ for all intralayer interactions. We see that $q$ is proportional to the weight of the directed link of the source node, and inversely proportional to the weighted total degree of the target node, a standard technique [163, 108]. We also vary the recovery rate, RR, which gives the probability that an infected node will recover at each time step. The simulation stops when there are no more infected nodes or when all nodes of the network have been infected.

## 8.4 Results

Initially, before we start the spreading process, we calculate the degree distribution of all layers and evaluate the slopes in the ensuing scale-free networks. The results are shown in Figure 8.2. We observe that the slopes are in the range of 1.06–1.42. Since the power-law exponent $\gamma$ is less than 2, the networks are very dense, and the average degree $< k >$ increases with the size of the network, so that the total number of links grows faster than the number of nodes [261, 294]. This implies that the addition of a link is inexpensive compared to the addition of a node [261]. Such networks are found in co-authorship in high energy physics, software package dependencies, word webs [261], and user-to-user links from the Facebook New Orleans (2009) networks [294].

We then start the spreading process. In Figure 8.3, we show a schematic of how the quantities of S, I, and R nodes behave as a function of time, for a specific set of parameter values, as shown in the figure, with the infection originating on a random node of the Low-Tech layer. We notice that, initially, all nodes are susceptible (fraction = 1.0), while no nodes are infected (fraction = 0). The number of recovered nodes starts at 0 and eventually reaches a constant value of 40% after about 200 Monte Carlo steps. We also see that, at all times, $S + I + R = 1$.

We performed 10,000 realizations; in Figure 8.4, the distribution of the infected mass $M$, that is, the percentage of nodes of a layer that were infected in a single realization, is plotted. Note that this distribution is bimodal [104], with one node being at very small values of the percentage of infected mass $M$, and the other one around $M \approx 85$.

In Figure 8.5, we plot the distribution of the infected mass for several different recovery rates, ranging from RR = 0.1 to RR = 0.25 for the case of no intra-urban connectivity. Note that that, for large values of RR, a decaying function ensues, while

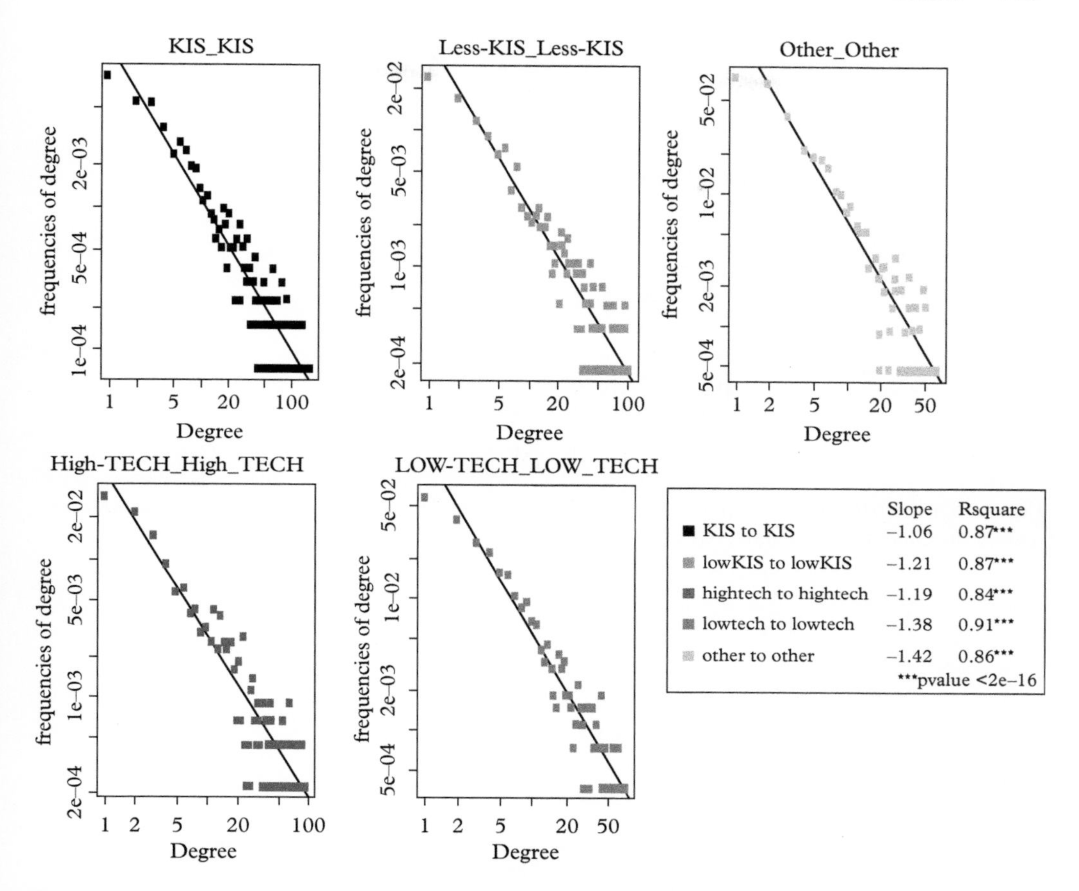


**Figure 8.2** *Degree distribution of the networks of the five layers; KIS, knowledge-intensive services.*

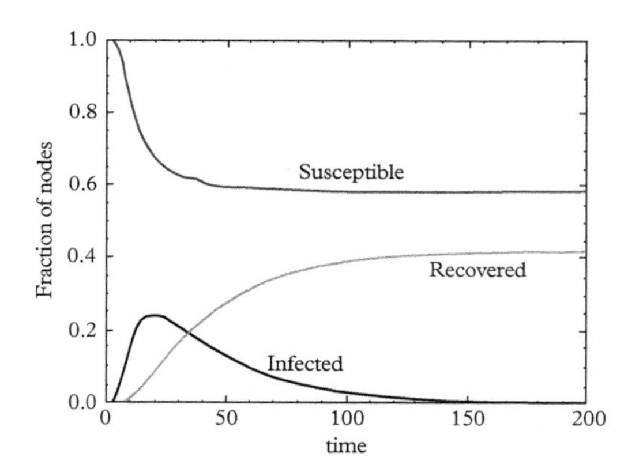


**Figure 8.3** *Evolution of the fractions of the S, I, and R nodes on the Low-Tech layer. The infection originated on a node in the Other layer; k =0.01, RR=1/30.*

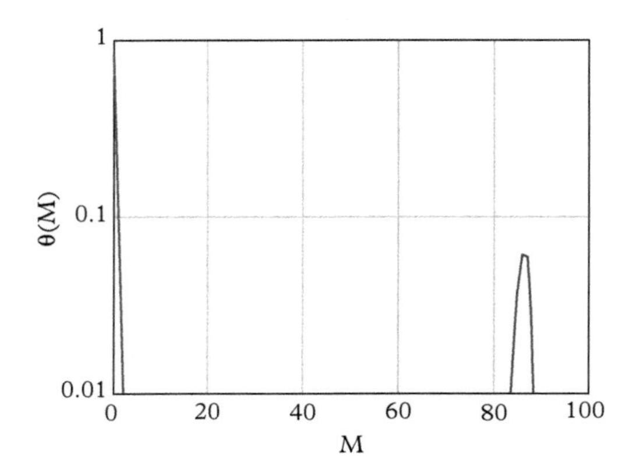


**Figure 8.4** *Distribution of the infected mass (M) of the Low-Tech layer. The infection originated on a node of the Other layer; k =0.01, RR=1/30.*

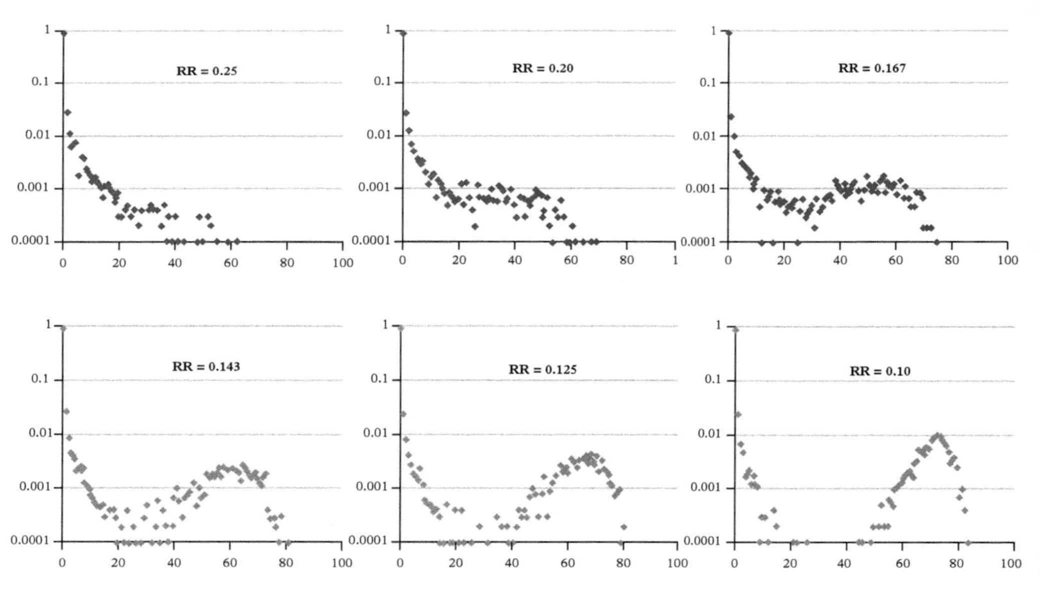


**Figure 8.5** *Distribution of the percentage of the infected mass of the Other layer, with the infection originating on a node of the same layer for k =0.01, and different RRs. The results are the average of 10,000 simulations.*

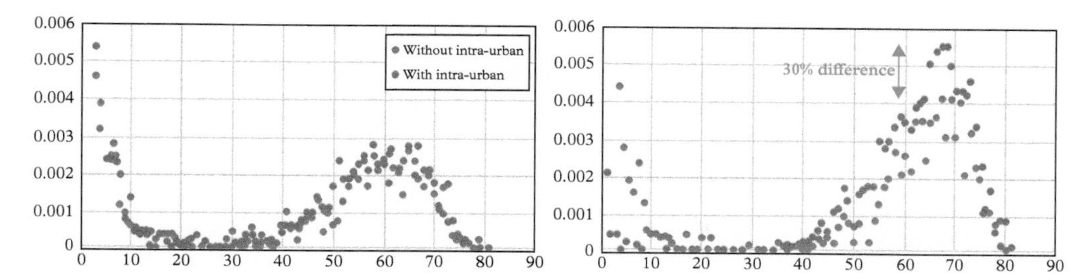


**Figure 8.6** *Distribution of the percentage of the infected mass of the Other layer, with the infection originating on a node of the same layer for k =0.01, and RR = 0.14 (left) and RR = 0.125 (right). Blue points are without intra-urban connections, and red points are with intra-urban connections.*

as RR decreases, the curve has a peak in large $M$ values, eventually becoming bimodal. Similar behavior was observed for all layers and starting nodes of infection.

In Figure 8.6, we add the results of simulations with intra-urban connections. No major difference is observed, within statistical error. The only difference of the order of 30%, is observed at large $M$ values and small RR ($M \approx 0.7$ and RR $\approx 0.125$). In this case, the intra-urban connections intensify significantly the scope of the spread, which corresponds to the process that we mentioned earlier when discussing the cascading failure of services in London during the 2008 crisis.

We now examine the time evolution of the fraction of infected nodes of the five layers, with the infection starting at different points of origin. We monitor the case for an origin at a specific layer of the five layers we have; in addition, we look at a random point of origin, and, finally, it the case of a single layer, that is, when there is no interlayer connectivity ($k = 0.0$) (see Figures 8.7, 8.8).

## 8.5 Conclusions

In all cases examined, we found a high probability of the infection dying out fast, that is, less than 1% of the nodes being infected. The reason that this happens is that, in scale-free networks, there are many nodes with only one (or very few) neighbors that can spread the crisis. Since the network is directed, many connected nodes cannot transmit the crisis. Finally, there is a probability that the crisis will not start on the largest connected component but on a small cluster, thus being confined in a very small space. However, if the crisis manages to survive the early time period, then it can propagate throughout the network. In such a case, we observe that the more the transmission is facilitated (large $k$, low RR), the stronger the bimodal effect is. In such cases, the crisis covers a large portion of the network, with an emergence of a gap in the infected mass ($M$) distribution, that is, there is almost zero probability that the crisis will cover an intermediate portion of the

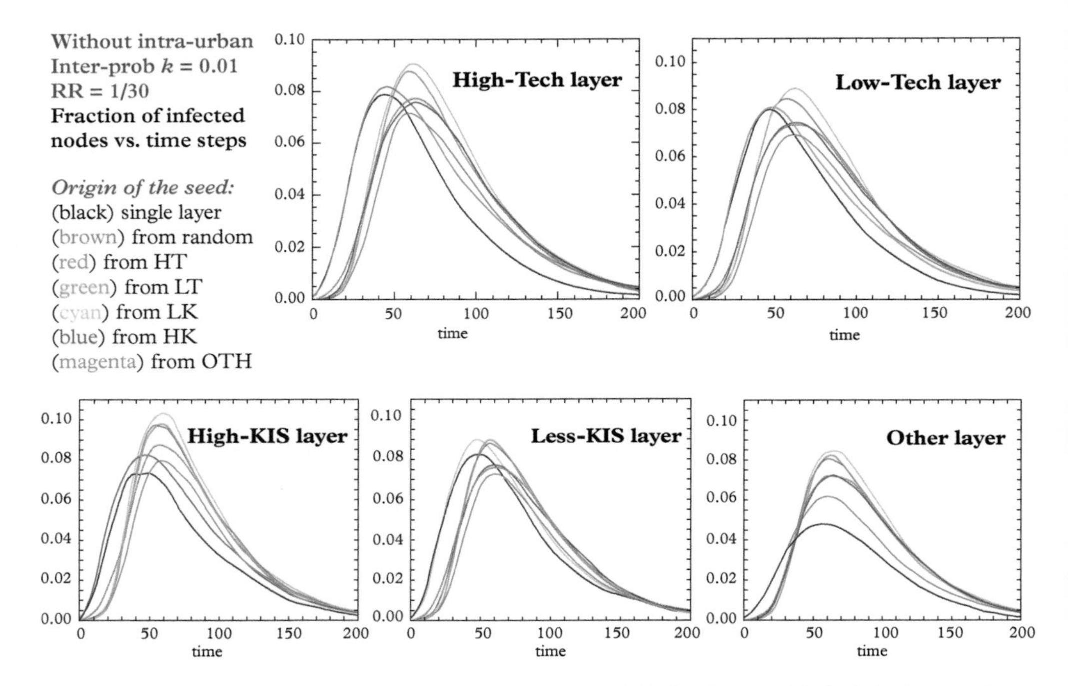


**Figure 8.7** *Time evolution of the fraction of infected nodes of the five layers, with the infection starting at different points of origin. The infection originates on a node of one of the five layers (colors: red, green, cyan, blue, and magenta), on a random node of any layer (brown color), or on a node of the same layer with no interlayer connectivity (k = 0.0, black color). The network does not include the intra-urban connections; HK, High-KIS layer; HT, High-Tech; LK, Less-KIS; LT, Low-Tech; OTH, other.*

network, and the infection will either stop at the very first nodes or cover almost all of the network.

Observing that the intra-urban connections intensify significantly the scope of the spread when RR is low demonstrates to what extent cities are vulnerable, in particular in times of crises with a low recovery rate. When the transmission is hampered (low $k$, large RR), then there is a finite probability that the crisis will cover intermediate portions of the network, and the intra-urban networks matter much less. This may indicate that hampering the diffusion of the crisis would be a was of supporting cities' sustainability. However, we must qualify this finding because when comparing the time evolution of the spreading of the crisis with and without intra-urban connections, we observe only small differences, in both the speed and the peak of the infection spreads, and these are are within the statistical error (see Figures 8.7 and 8.8).

Moreover, another major result of this model is the evidence that, in all cases, the maximum intensity of the affected population M is seen when the crisis starts at the less-KIS layer. These less-KIS activities neither are the most numerous nor constitute

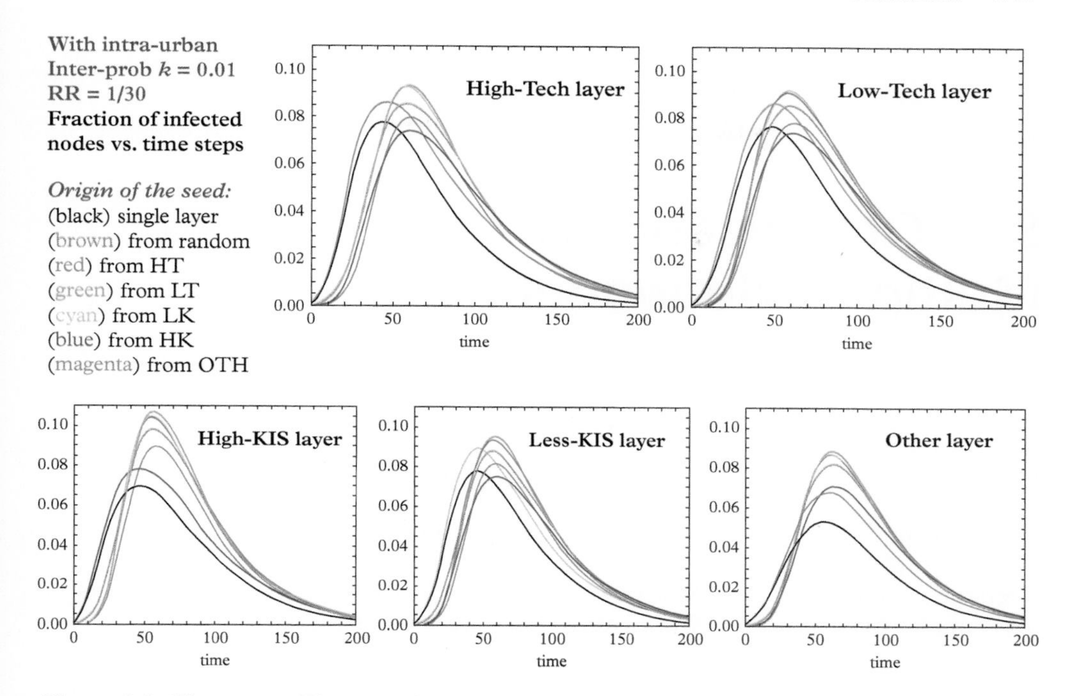


**Figure 8.8** *The same as Figure 8.7 but including the intra-urban connections; HK, High-KIS layer; HT, High-Tech; LK, Less-KIS; LT, Low-Tech; OTH, other.*

the most central layer (exchanging more with other layers in the enterprises networks); in fact, the high-KIS activities are much more dominant in this perspective. So, we could wonder if some specific positioning of less-KIS activities in the network of firms would create this phenomenon of the spreading amplification. All the basic structural network indices that we have evaluated to this point have not answered this question.

# 9

# Self-Organization in Multiplex Networks

**Nikos E. Kouvaris**[1] **and Albert Díaz-Guilera**[2]

[1]Department of Mathematics, Namur Institute for Complex Systems (naXys), University of Namur, Namur, Belgium
[2]Departament de Fisica de la Materia Condensada, Universitat de Barcelona Institute of Complex Systems (UBICS), Universitat de Barcelona, Barcelona, Spain

A great question in the research of complex systems concerns the way the network structure shapes the hosted dynamics and leads to a plethora of self-organization phenomena. Complex systems consist of nodes having some intrinsic dynamics, usually nonlinear, and are connected through the links of the network. Such systems can be studied by means of discrete reaction–diffusion equations; reaction terms account for the dynamics in the nodes whereas diffusion terms describe the coupling between them. Here, we discuss how multiplex networks are suitable for studying such systems, by providing two illustrative examples of self-organization phenomena occurring in them.

First, we deal with synchronization phenomena in a network originated from the neural connectome of the *Caenorhabditis elegans* nematode. In this network, neurons can be connected with two types of synapses: electrical and chemical. Therefore, the network can be treated as a multiplex where different types of synapses are represented in different networks that correspond to distinct layers. Second, we employ the framework of multiplex networks in distributed ecosystem networks, where different species migrate along different paths. In the latter case, an interesting question arises concerning how the multiplex structure can induce instabilities leading to Turing patterns that represent distributed patches of the two species.

## 9.1 The multiplex structure and the dynamics of neural networks

Here, we consider the neural network of the *C. elegans* nematode, which has a modular structure [301] and is equipped with electrical and chemical synapses for communication

Kouvaris N. E. and Díaz-Guilera, A., "Self-Organization in Multiplex Networks" in *Multiplex and Multilevel Networks*, edited by Battiston, S., Caldarelli, G., and Garas, A. © Oxford University Press 2019.
DOI: 10.1093/oso/9780198809456.003.0009

[13, 131]. We also assume that neurons obey chaotic bursting dynamics and are connected via electrical synapses within their communities, and chemical synapses across them. The coaction of those synapses results in the emergence of synchronized states or chimera states.

An essential question in theoretical neuroscience concerns the relationship between *structure* and *function*, namely, how the structure of the synapses connecting different neurons or regions of the brain can shape its dynamical state. A synapse is a junction between two neurons that allows them to communicate with each other. Specifically, electrical synapses are physical connections that allow electrons to pass through neurons by very small gaps between them; they are bidirectional and connect neurons within their vicinity. Chemical synapses are typically unidirectional, and the presynaptic signals are transmitted via the release of neurotransmitters from the presynaptic neuron, which attaches to receptors at the postsynaptic neuron; they can connect neurons that are close to or far from each other.

For the purpose of this discussion, let us employ a community detection method and split the *C. elegans* connectome into six interconnected communities [131]. Based on the detected communities, we assume that neurons within the same community are connected with electrical synapses, and neurons across communities with chemical synapses. For the sake of simplicity, we assume only bidirectional synapses. Such a network can naturally be seen as a multiplex network, where the two types of synapses are organized into two distinct layers (see Figure 9.1).

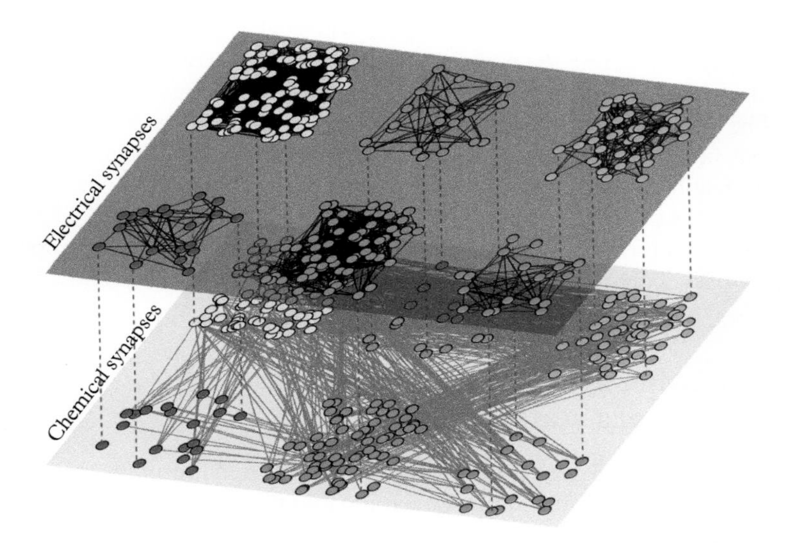


**Figure 9.1** *The multiplex structure of the connectome of the C. elegans neural network. Neurons are organized into six communities and are shown in different color according to the community to which they belong. Different types of synapses are placed in two distinct layers; electrical synapses connect neurons within the same community, whereas chemical synapses across different communities.*

In the following, we assume that the neurons perform chaotic bursting oscillations described by the Hindmarsh–Rose model. Therefore, the evolution of the multiplex network where the neurons are connected by electrical and chemical synapses is described by the equations

$$\begin{aligned}
\dot{p}_i &= q_i - ap_i^3 + bp_i^2 - n_i + I_{\text{ext}} + \underbrace{g_{el} \sum_{j=1}^{N} L_{ij} H(p_j)}_{\text{term for electrical synapses}} \underbrace{- g_{ch}(p_i - V_{\text{syn}}) \sum_{j=1}^{N} T_{ij} S(p_j)}_{\text{term for chemical synapses}}, \\
\dot{q}_i &= c - dp_i^2 - q_i, \\
\dot{n}_i &= r[s(p_i - p_0) - n_i], \qquad (9.1)
\end{aligned}$$

where $i = 1, \ldots, N$ is the neuron index, $p_i$ is the membrane potential of the $i$th neuron, $q_i$ is associated with the fast current (either $Na^+$ or $K^+$), and $n_i$ is associated with the slow current (e.g., $Ca^{2+}$). The parameters of the equations in (9.1) are chosen such that $a = 1$, $b = 3$, $c = 1$, $d = 5$, $s = 4$, $p_0 = -1.6$, and $I_{\text{ext}} = 3.25$, for which the system exhibits a chaotic bursting behavior; $r$ characterizes the slow dynamics of the system; for $r = 0.005$ all neurons perform multiscale chaotic oscillations. For these parameters, the Hindmarsh–Rose model enables the spiking–bursting behavior of the membrane potential observed in experiments made with single neurons in vitro.

Electrical synapses support linear coupling and thus are described in terms of the Laplacian matrix $\mathbf{L}$, whose elements are defined as $L_{ij} = E_{ij} - \delta_{ij} k_i$, where $\delta_{ij} = 1$ if $i = j$, and $\delta_{ij} = 0$ otherwise; $\mathbf{E}$ is an adjacency matrix whose elements are $E_{ij} = 1$ if there is an electrical synapse connecting the neurons $i$ and $j$, and $E_{ij} = 0$ otherwise. The strength of the electrical synapses is given by the parameter $g_{el}$ and their functionality by the kernel $H(p) = p$. The connectivity pattern of the chemical synapses is described in terms of the adjacency matrix $\mathbf{T}$, whose elements are $T_{ij} = 1$ if there is a chemical synapse between neurons $i$ and $j$, and $T_{ij} = 0$ otherwise. Chemical synapses carry nonlinear coupling, and their functionality is described by the sigmoidal kernel $S(p) = \{1 + \exp[-\lambda(p - \theta_{\text{syn}})]\}^{-1}$, which acts as a continuous mechanism for the potentiation and depression of the chemical synapses. Chemical synapses have strength $g_{ch}$ and are assumed here to be excitatory ($V_{\text{syn}} = 2$). The other parameters are $\theta_{\text{syn}} = -0.25$ and $\lambda = 10$.

### 9.1.1 Estimating synchronization, metastability, and chimera states

Networks of coupled oscillators usually can reach stable synchronized states. Nevertheless, they can also be metastable in time, meaning that they can stay in the vicinity of one stable state for a certain time interval and then, spontaneously, move toward another. An even more interesting feature of many complex systems is, undoubtedly, the coexistence of different, often contradictory states. In terms of synchronization, this is illustrated through the so-called *chimera* states [164, 1], where one population of oscillators synchronizes whereas other populations of identical oscillators are desynchronized. Below, we provide three measures [263] which characterize each of the aforementioned states for multiplex (or modular) networks such the one discussed here.

#### 9.1.1.1 The Kuramoto order parameter

The order parameter $\rho$ is employed to account for the synchronization level of the neural activity of the considered networks and of their communities [120]. It originated from the theory of measures of dynamical coherence of a population of $N$ Kuramoto phase oscillators [164] and can be computed by a complex number $z(t)$ defined as

$$z(t) = \rho(t)e^{\mathrm{i}\Phi(t)} = \frac{1}{N}\sum_{j=1}^{N} e^{\mathrm{i}\phi_j(t)}. \tag{9.2}$$

By taking the modulus $\rho(t)$ of $z(t)$, one can measure the phase coherence of the $N$ neurons of the network, and, by $\Phi(t)$, their average phase. In this context, the phase of the $j$th neuron at time $t$ is given by $\phi_j(t) = \arctan(q_j(t)/p_j(t))$.

#### 9.1.1.2 The metastability index

The metastability and chimera-likeness of the observed dynamics can be characterized by two measures first introduced by M. Shanahan [263]. Specifically, the level of metastability can be calculated from the index $\lambda$, where

$$\lambda = \langle \sigma_{\mathbf{met}} \rangle_{C_m}, \tag{9.3}$$

and

$$\sigma_{\mathbf{met}}(m) = \frac{1}{T-1}\sum_{t=1}^{T}(\rho_m(t) - \langle \rho_m \rangle_T)^2. \tag{9.4}$$

In Eq. (9.3), $C_m$ is the set of all $M$ communities, and $m = 1,2,\ldots,M$. This is the number of the disconnected networks in the electrical layer (see Figure 9.1). The order parameter $\rho_m(t)$ of each community $m$ is sampled at discrete times $t \in 1,\ldots,T$. The variance $\sigma_{\mathbf{met}}(m)$ of $\rho_m(t)$ over time gives an indication of how much the synchrony in this community fluctuates in time. Obviously, by averaging over all communities, we can estimate the metastability for the entire network.

#### 9.1.1.3 The chimera-like index

Similarly, the chimera-like index $\chi$ [263] reads

$$\chi = \langle \sigma_{\mathbf{chi}} \rangle_T, \tag{9.5}$$

where

$$\sigma_{\mathbf{chi}}(t) = \frac{1}{M-1}\sum_{m=1}^{M}(\rho_m(t) - \langle \rho(t)\rangle_M)^2; \tag{9.6}$$

$\sigma_{\mathbf{chi}}(t)$ is an instantaneous quantity that gives the variance of $p_m(t)$ over all $M$ communities (i.e., the number of the disconnected networks in the electrical layer shown in Figure 9.1) at a given time $t$. By averaging this quantity in time, we estimate how much the synchrony in each community differs from the synchrony in the other communities.

The density plots in Figure 9.2 show the values of $\rho$, $\lambda$, and $\chi$ for a range of the electrical and chemical coupling $g_{el}$ and $g_{ch}$, respectively. We select three points of interest (A, B, and C) on the parameter space $g_{ch} - g_{el}$ and highlight the most interesting self-organized behaviors that emerge in this multiplex dynamical network.

Point $A$ corresponds to low values of $\lambda$ and $\chi$. This means that the network does not visit frequently different synchronization states, and the six communities are, to a large extent, simultaneously in synchrony with each other. This is expected for such a combination of electrical and chemical couplings and is in agreement with the high value of $\rho$. Figure 9.3(a) illustrates clearly this synchronous steady state. The corresponding time series exhibits a quite regular spiking behavior.

Point $B$ shows the values of $g_{el}$ and $g_{ch}$, where $\rho$ and $\chi$ are low whereas $\lambda$ is high. Therefore, it is expected that the network dynamics will be metastable. Indeed, this is illustrated in Figure 9.3(b) by the rather regular pattern in space which, in time, switches between slow, quiescent periods to fast-spiking intervals that correspond to synchronous and incoherent regimes, respectively.

An interesting behavior is found at the point $C$, which corresponds to moderate values of $\rho$ but low $\lambda$ and high $\chi$. The corresponding time series shows that the system dynamics lies in the bursting regime. Point $C$ shows the values of $g_{el}$ and $g_{ch}$, where $\rho$ has a moderate value and lies at the edge between the synchronous and asynchronous regimes but $\chi$ is high whereas $\lambda$ is low. This point corresponds to a chimera-like state where, as Figure 9.3(c) shows, communities 1, 3, and 5 seem to be quite incoherent, whereas

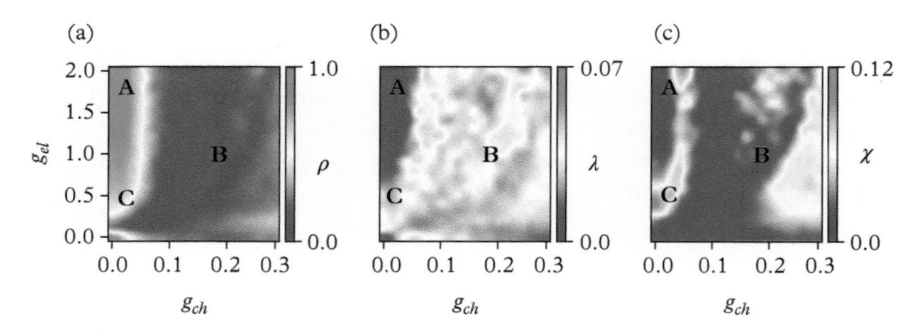


**Figure 9.2** *Parameter spaces in the plane $g_{ch} - g_{el}$ of (a) the global order parameter $\rho$, (b) the metastability index $\lambda$, and (c) the chimera-like index $\chi$.*

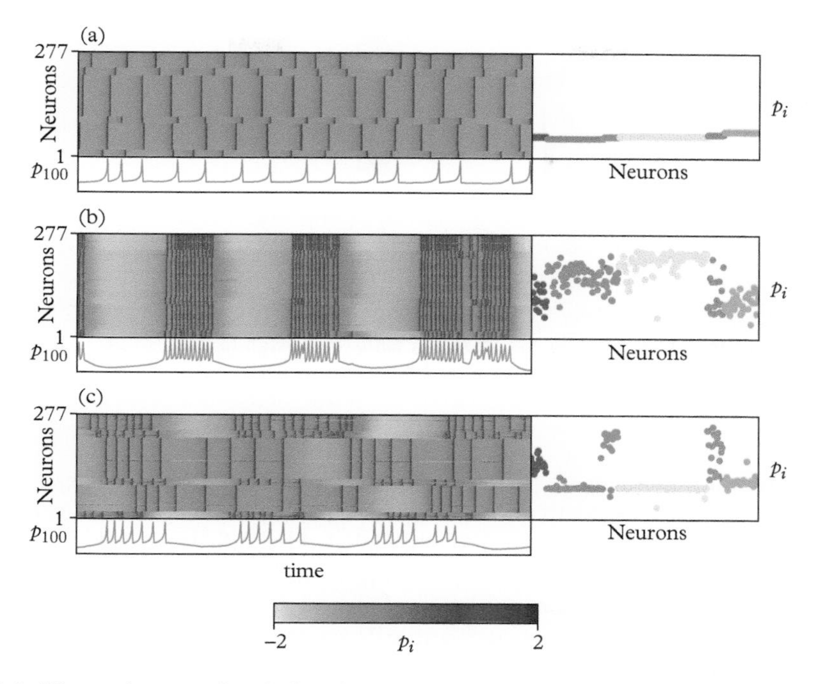


**Figure 9.3** *The spatiotemporal evolution of $p_i$ (upper left), with a time series of the neuron with index 100 of community 3 (down left) and a snapshot of the system's state (right) are shown for the values of $g_{el}$ and $g_{ch}$ corresponding to the points (a) A, (b) B, and (c) C.*

communities 2, 4, and 6 show long synchronization intervals. The corresponding snapshots, where coexisting synchronized and unsynchronized communities are clearly presented, resembles the well-known chimera states reported in many systems both theoretically and experimentally [224].

## 9.2 Activator–inhibitor dynamics in multiplex networks

Now we consider a different example of multiplex dynamical networks where two interacting populations, the activators and the inhibitors, occupy separate network nodes in distinct layers [157]. They react across layers according to the mechanism defined by the activator–inhibitor dynamics and migrate to other nodes in their own layer through the connecting links (see Figure 9.4). Such a process can be described by the equations

$$\frac{d}{dt}u_i(t) = f(u_i, v_i) + \sigma^{(u)} \sum_{j=1}^{N} L_{ij}^{(u)} u_j, \tag{9.7a}$$

$$\frac{d}{dt}v_i(t) = g(u_i, v_i) + \sigma^{(v)} \sum_{j=1}^{N} L_{ij}^{(v)} v_j, \tag{9.7b}$$

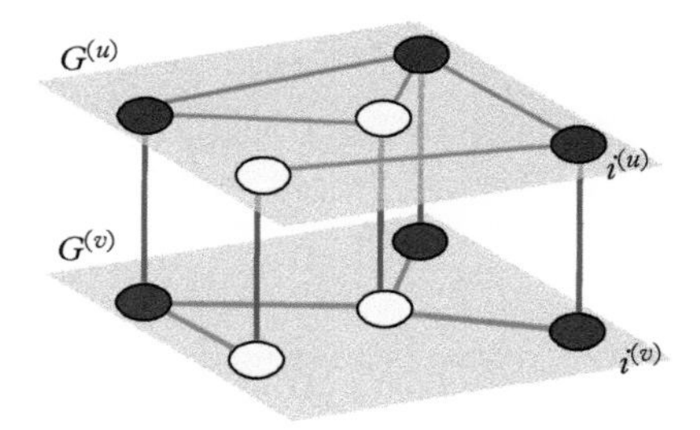


**Figure 9.4** *Activator and inhibitor species occupy nodes in separate layers $G^{(u)}$ and $G^{(v)}$, respectively. They react across the layers (blue interlayer links), while they migrate within their own layers (green intralayer links).*

where $u_i$ and $v_i$ are the densities of activator and inhibitor species in nodes $i^{(u)}$ and $i^{(v)}$ of layers $G^{(u)}$ and $G^{(v)}$, respectively. The superscripts $(u)$ and $(v)$ refer to activator and inhibitor, respectively. The activator nodes are labeled by indices $i = 1, 2, \ldots, N$, in order of decreasing connectivity. The functions $f(u_i, v_i)$ and $g(u_i, v_i)$ specify the activator–inhibitor dynamics. The Laplacian matrices $L^{(u)}$ and $L^{(v)}$ describe diffusion processes in the two layers, and the constants $\sigma^{(u)}$ and $\sigma^{(v)}$ are the corresponding mobility rates.

As a particular example, we consider the Mimura–Murray ecological model [205] on a multiplex network consisting of two layers [157, 14]. Its dynamics is given by the functions

$$f(u,v) = \left( \frac{a + bu - u^2}{c} - v \right) u,$$
$$g(u,v) = (u - dv - 1)v, \tag{9.8}$$

where $u, v$ correspond to the densities of the activator and the inhibitor, respectively. In the absence of diffusive coupling ($\sigma^{(u)} = 0$, and $\sigma^{(v)} = 0$), the multiplex system relaxes to a uniform steady state $(u_i, v_i) = (u_0, v_0)$ for all $i = 1, \ldots, N$. The homogeneous densities are determined by $f(u_0, v_0) = g(u_0, v_0) = 0$. This requires the networks to satisfy $(\mathcal{J}_{(u_0,v_0)}) < 0$ and $\det(\mathcal{J}_{(u_0,v_0)}) > 0$, where $\mathcal{J}$ is the Jacobian matrix

$$\mathcal{J}_{(u,v)} = \begin{pmatrix} f_u & f_v \\ g_u & g_v \end{pmatrix}, \tag{9.9}$$

and $f_{u,v}$ and $g_{u,v}$ denote the partial derivatives. In simplex networks, where $L^{(u)} \equiv L^{(v)}$, the uniform state may undergo a Turing instability as the ratio $\sigma^{(v)}/\sigma^{(u)}$ increases and exceeds a certain threshold. The instability leads to the spontaneous emergence of stationary patterns consisting of nodes with high or low densities of activators [205]. Such diffusion-induced instability can also take place in multiplex reaction networks (9.7). This can be explained through a linear stability analysis with nonuniform perturbations.

### 9.2.1 Linear stability analysis

We introduce small perturbations $(\delta u_i, \delta v_i)$ to the uniform steady state $(u_0, v_0)$, as $(u_i, v_i) = (u_0, v_0) + (\delta u_i, \delta v_i)$. Substituting into Eqs (9.7a) and (9.7b), we obtain the linearized differential equations

$$\frac{d}{dt}\delta u_i = f_u \delta u_i + f_v \delta v_i + \sigma^{(u)} \sum_{j=1}^{N} L_{ij}^{(u)} \delta u_j, \tag{9.10}$$

$$\frac{d}{dt}\delta v_i = g_u \delta u_i + g_v \delta v_i + \sigma^{(v)} \sum_{j=1}^{N} L_{ij}^{(v)} \delta v_j. \tag{9.11}$$

Alternatively, the linearized differential equations can be written as

$$\frac{d}{dt}\mathbf{w} = (\mathcal{J} + \mathcal{L})\mathbf{w}, \tag{9.12}$$

where $\mathbf{w} = (\delta u_1, \cdots, \delta u_N, \delta v_1, \cdots, \delta v_N)^T$ is the perturbation vector,

$$\mathcal{J}_{(u,v)} = \begin{pmatrix} f_u I & f_v I \\ g_u I & g_v I \end{pmatrix} \tag{9.13}$$

is the supra-Jacobian matrix,

$$\mathcal{L} = \begin{pmatrix} \sigma^{(u)} L^{(u)} & 0 \\ 0 & \sigma^{(v)} L^{(v)} \end{pmatrix} \tag{9.14}$$

is the supra-Laplacian matrix, and $I$ is the $N \times N$ identity matrix. For the linear stability analysis, the perturbation vector $\mathbf{w}$ should be expanded over the set of eigenvectors of the matrix $\mathcal{Q} = \mathcal{J} + \mathcal{L}$. However, we cannot calculate them for different network topologies, that is, different Laplacian matrices $L^{(u)}$ and $L^{(v)}$. Therefore, we propose an approximation technique to analyze the linear stability of the system. We split the supra-Laplacian matrix $\mathcal{L}$ into $\mathcal{L} = \mathcal{Q}_0 - \mathcal{D}$, where

$$\mathcal{Q}_0 = \begin{pmatrix} \sigma^{(u)} A^{(u)} & 0 \\ 0 & \sigma^{(v)} A^{(v)} \end{pmatrix} \tag{9.15}$$

and

$$\mathcal{D} = \begin{pmatrix} \sigma^{(u)} D^{(u)} & 0 \\ 0 & \sigma^{(v)} D^{(v)} \end{pmatrix}. \tag{9.16}$$

The matrices $A^{(u)}$ and $A^{(v)}$ are the adjacency matrices of layers $G^{(u)}$ and $G^{(v)}$, respectively. The matrices $D^{(u)}$ and $D^{(v)}$ are the corresponding degree matrices, which have the nodes degrees in the main diagonal and are zero elsewhere. Then, the matrix $\mathcal{Q}$ can be rewritten as $\mathcal{Q} = \mathcal{Q}_0 + \mathcal{Q}_1$, where

$$\mathcal{Q}_1 = \begin{pmatrix} f_u I - \sigma^{(u)} D^{(u)} & f_v I \\ g_u I & g_v I - \sigma^{(v)} D^{(v)} \end{pmatrix}. \tag{9.17}$$

Examining the matrices $\mathcal{Q}_0$ and $\mathcal{Q}_1$, we see that the first has elements with values of order $\mathcal{O}(\sigma^{(u)})$ or $\mathcal{O}(\sigma^{(v)})$, while the second has elements with values of order $\mathcal{O}(\sigma^{(u)}\langle k^{(u)}\rangle)$ or $\mathcal{O}(\sigma^{(v)}\langle k^{(v)}\rangle)$. If both layers are dense enough so that $\langle k^{(u)}\rangle \gg 1$ and $\langle k^{(v)}\rangle \gg 1$, we can clearly see that the elements of matrix $\mathcal{Q}_1$ have significantly larger values than those of matrix $\mathcal{Q}_0$ and thus $\mathcal{Q}_0$ can be neglected. This approximation yields the linearized equation

$$\frac{d}{dt}\mathbf{w} = \mathcal{Q}_1 \mathbf{w}, \tag{9.18}$$

from which we obtain the characteristic equation for the growth rate $\lambda$ of the perturbations for each pair of nodes,

$$\det \begin{pmatrix} f_u - \sigma^{(u)} k^{(u)} - \lambda & f_v \\ g_u & g_v - \sigma^{(v)} k^{(v)} - \lambda \end{pmatrix} = 0$$

and is the same for each pair of nodes $i^{(v)}, i^{(u)}$. This approximation neglects entirely the matrix $\mathcal{Q}_0$, which is associated with the precise architectures of the layers. Instead, each node is characterized only by its degree. This is quite similar to the powerful mean-field methods used for analyzing Turing patterns in single-layer networks [205, 312].

The instability occurs when $\mathrm{Re}\,\lambda = 0$. This condition is fulfilled when the pair of nodes $i^{(v)}$ and $i^{(u)}$ appears to have a combination of degrees $k^{(u)}$ and $k^{(v)}$ for which the equation

$$k^{(u)} = \frac{f_u g_v - f_v g_u - f_u \sigma^{(v)} k^{(v)}}{g_v \sigma^{(u)} - \sigma^{(u)} \sigma^{(v)} k^{(v)}} \tag{9.19}$$

is satisfied. We clearly see that a sufficiently large value of $\sigma^{(v)}$ brings about instability, similarly to the Turing instability. However, an alternative scenario of the instability is revealed by Eq. (9.19). This instability shows differences from the classical Turing instabilities and occurs by increasing $k^{(v)}$ even if the mobilities are equal ($\sigma^{(u)} = \sigma^{(v)}$).

Figure 9.5(a) shows the linear stability of system (9.7) for varying $k^{(v)}$, holding $k^{(u)}$ fixed. We clearly see that the uniform steady state is always a solution of the multiplex system. It is linearly stable (the green line) for small values of $k^{(v)}$; but at some critical value of $k^{(v)}$ which satisfies Eq. (9.19), the system undergoes a transcritical bifurcation (the red point) and becomes unstable (the magenta line). Two new branches of solutions arise from the transcritical bifurcation. The unstable branch (the magenta line) undergoes a second bifurcation (the blue point), this time a saddle-node bifurcation, giving rise to a new branch of stable solutions (the green line) different from the uniform steady state.

Figure 9.5(b) shows the two bifurcations in the $k^{(v)}$–$k^{(u)}$ plane. Based on the latter figure, we describe the bifurcation scenario as follows. Let us consider a multiplex network whose dynamics starts almost in the uniform steady state with a small perturbation. By examining Eq. (9.19), we can identify pairs of nodes $(i^{(v)}, i^{(u)})$ where the

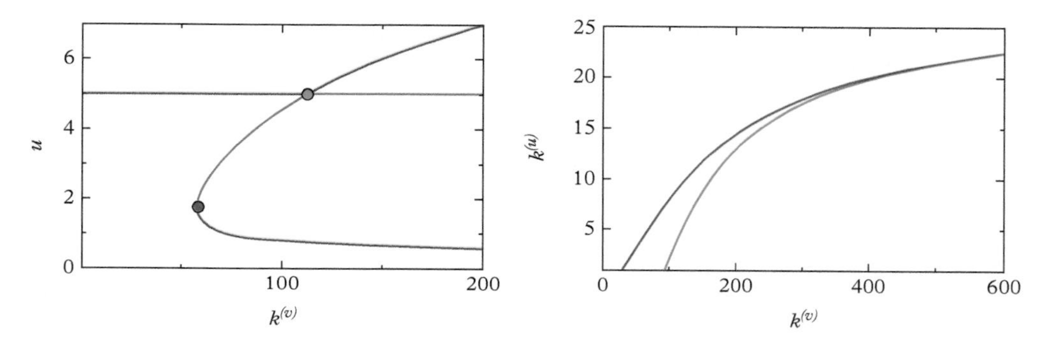


**Figure 9.5** *Bifurcation diagram. (Left) Stationary solutions of system (9.7) for $k^{(u)} = 4$. Green curves indicate stable solutions, while magenta curves correspond to unstable solutions of the linearized system. The red point indicates the transcritical bifurcation where the uniform steady state $(u_0, v_0) = (5,10)$ becomes unstable. The blue point corresponds to a saddle-node bifurcation of a solution $(u, v)$ which originates from the transcritical bifurcation. (Right) Transcritical bifurcation (the red curve) given by Eq. (9.19) is shown, together with the continuation of the saddle-node bifurcation (the blue curve) in the plane $k^{(v)}$–$k^{(u)}$.*

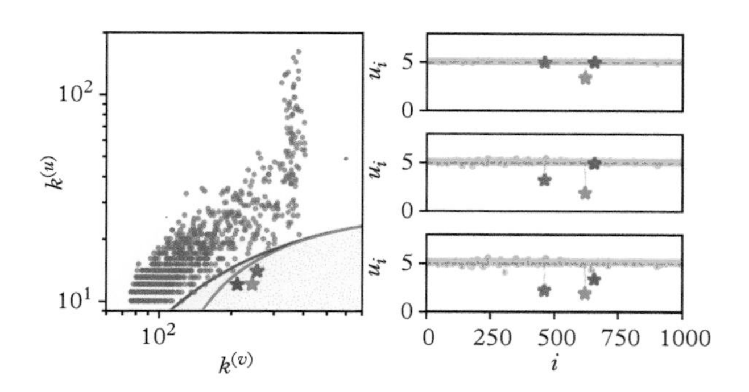


**Figure 9.6** *Diffusion-induced instability. (Left) Degree combination for pairs of nodes $i^{(v)}$ and $i^{(u)}$ is shown in plane $k^{(v)}$–$k^{(u)}$, together with the curves of saddle-node (blue) and transcritical (red) bifurcations. (Right) Snapshots of the activator pattern for $t = 50$, $t = 63$, $t = 70$, and the fully developed pattern are shown for the Mimura–Murray model with $\sigma^{(v)} = \sigma^{(u)} = 0.12$ on a multiplex network with scale-free layers of $N = 1{,}000$ nodes and mean degrees $\langle k^{(v)} \rangle = 152$ and $\langle k^{(u)} \rangle = 20$. Nodes are ordered according to decreasing degrees $k^{(u)}$. See the text for further details.*

small perturbations will be amplified faster. These pairs of nodes are the first to leave the uniform state and trigger the formation of a multiplex-organized Turing pattern. Numerical simulations can be used to verify this scenario. The left panel of Figure 9.6 shows the actual degree combination $(k^{(v)}, k^{(u)})$ for all pairs of nodes $i^{(v)}, i^{(u)}$ (blue dots) of a multiplex network in the $k^{(v)}$–$k^{(u)}$ plane, together with the bifurcation curves. One can easily detect three pairs of nodes, denoted by the star shape in the figure, which have degrees exceeding the instability threshold. We expect that a Turing-like pattern will be formed starting from these nodes. The critical node shown by the red star is the first to spontaneously leave the uniform state, as shown in the left panel of Figure 9.6. Next, the critical nodes denoted by the green and blue stars also differentiate from the uniform state. Finally, triggered by these growing perturbations, most of the nodes leave the steady state and establish a nonuniform pattern.

All pairs of nodes lying after the transcritical bifurcation differentiate from the steady state and lead to the formation of the Turing patterns, as described above. However, the pairs of nodes with degrees before the transcritical but after the saddle-node bifurcation can also trigger the Turing-like instability after aplying stronger perturbations. This indicates that the system exhibits multistability between the two bifurcations, where the uniform steady state coexists with a branch of solutions corresponding to nonuniform patterns.

# References


[1] Abrams, D. M., and Strogatz, S. H. (2004). Chimera states for coupled oscillators. *Physical Review Letters*, 93(17):174102.

[2] Coolen, A. C. C., De Martino, A., and Annibale, A. (2009). Constrained Markovian dynamics of random graphs. *Journal of Statistical Physics*, 136(6):1035–67.

[3] Ahn, Y.-Y., Bagrow, J. P., and Lehmann, S. (2010). Link communities reveal multiscale complexity in networks. *Nature*, 466(7307):761–4.

[4] Aizenman, J., Hutchison, M., and Jinjarak, Y. (2013). What is the risk of European sovereign debt defaults? Fiscal space, CDS spreads and market pricing of risk. *Journal of International Money and Finance*, 34(C):37–59.

[5] Albert, R., and Barabási, A.-L. (2002). Statistical mechanics of complex networks. *Reviews of Modern Physics*, 74(1): 47.

[6] Alderson, A. S., and Beckfield, J. (2004). Power and position in the world city system. *American Journal of Sociology*, 109(4):811–51.

[7] Almeida-Neto, M., Guimarães, P., Guimarães, P. R., Loyola, R. D., and Ulrich, W. (2008). A consistent metric for nestedness analysis in ecological systems: reconciling concept and measurement. *Oikos*, 117(8):1227–39.

[8] Alon, N., Caro, Y., Krasikov, I., and Roditty, Y. (1989). Combinatorial reconstruction problems. *Journal of Combinatorial Theory Series B*, 47:153–61.

[9] Alvarez-Hamelin, J., Dall'Asta, L., Barrat, A., and Vespignani, A. (2008). K-core decomposition of Internet graphs: Hierarchies, self-similarity and measurement biases. *Networks and Heterogeneous Media*, 3(2):371–93.

[10] Ángeles Serrano, M., and Boguñá, M. (2005). Tuning clustering in random networks with arbitrary degree distributions. *Physical Review E*, 72(3):36133.

[11] Annibale, A., Coolen, A. C. C., Fernandes, L., Fraternali, F., and Kleinjung, J. (2009). Tailored graph ensembles as proxies or null models for real networks I: tools for quantifying structure. *Journal of Physics A, Mathematical and General*, 42(48):485001.

[12] Anthonisse, J. M. (1971). The rush in a directed graph. *Stichting Mathematisch Centrum. Mathematische Besliskunde*, (BN 9/71):1–10.

[13] Antonopoulos, C. G., Srivastava, S., Pinto, S. E. S., and Baptista, M. S. (2015). Do brain networks evolve by maximizing their information flow capacity? *PLoS Computational Biology*, 11(8):e1004372.

[14] Asllani, M., Busiello, D. M., Carletti, T., Fanelli, D., and Planchon, G. (2014). Turing patterns in multiplex networks. *Physical Review E*, 90(4):042814.

[15] Barabási, A.-L. (2016). *Network Science*. Cambridge University Press, Cambridge.

[16] Barabási, A.-L., and Albert, R. R. (1999). Emergence of scaling in random networks. *Science*, 286(5439):509–12.

[17] Barrat, A., Barthélemy, M., and Vespignani, A. (2008). *Dynamical Processes on Complex Networks*. Cambridge University Press, Cambridge.

[18] Barrat, A., Cattuto, C., Tozzi, A., Vanhems, P., and Voirin, N. (2014). Measuring contact patterns with wearable sensors: Methods, data characteristics and applications to data-driven simulations of infectious diseases. *Clinical Microbiology and Infection*, 20(1):10–16.

[19] Barthélemy, M. (2011). Spatial networks. *Physics Reports*, 499(1–3):1–101.
[20] Barvinok, A., and Hartigan, J. A. (2013). The number of graphs and a random graph with a given degree sequence. *Random Structures and Algorithms*, 42(3):301–48.
[21] Bascompte, J., Jordano, P., Melián, C. J., and Olesen, J. M. (2003). The nested assembly of plant–animal mutualistic networks. *Proceedings of the National Academy of Sciences of the United States of America*, 100(16):9383–7.
[22] Bassler, K. E., Del Genio, C. I., Erdős, P. L., Miklós, I., and Toroczkai, Z. (2015). Exact sampling of graphs with prescribed degree correlations. *New Journal of Physics*, 17(8): 083052.
[23] Bastian, M., Heymann, S., and Jacomy, M. (2009). Gephi: An open source software for exploring and manipulating networks. In Adar, E., Hurst, M., Finin, T., Glance, N. S., Nicolov, N., and Tseng, B. L. (eds), *Proceedings of the Third International Conference on Weblogs and Social Media, ICWSM 2009, San Jose, California, USA, May 17–20*, pp. 361–2. AAAI Press, Palo Alto.
[24] Bastolla, U., Fortuna, M. A., Pascual-Garcia, A., Ferrera, A., Luque, B., and Bascompte, J. (2009). The architecture of mutualistic networks minimizes competition and increases biodiversity. *Nature*, 458(7241):1018–20.
[25] Batagelj, V., and Mrvar, A. (2004). Pajek: Analysis and visualization of large networks. In Jünger, M., and Mutzel, P. (eds), *Graph Drawing Software*, pp. 77–103. Springer, Berlin, Heidelberg.
[26] Battiston, F., Iacovacci, J., Nicosia, V., Bianconi, G., and Latora, V. (2016). Emergence of multiplex communities in collaboration networks. *PLOS ONE*, 11(1):e0147451.
[27] Battiston, F., Nicosia, V., and Latora, V. (2014). Structural measures for multiplex networks. *Physical Review E*, 89(3):32804.
[28] Bavelas, A. (1950). Communication patterns in task-oriented groups. *The Journal of the Acoustical Society of America*, 22(6):725–30.
[29] Beiró, M. G., Alvarez-Hamelin, J. I., and Busch, J. R. (2008). A low complexity visualization tool that helps to perform complex systems analysis. *New Journal of Physics*, 10(12):125003.
[30] Bernard, H. R. (2000). *Social Research Methods: Qualitative And Quantitative Approaches.* Sage Publications, Los Angelos.
[31] Bettencourt, L. M., Lobo, J., Helbing, D., Kühnert, C., and West, G. B. (2007). Growth, innovation, scaling, and the pace of life in cities. *Proceedings of the National Academy of Sciences of the United States of America*, 104(17):7301–6.
[32] Bianconi, G. (2009). Entropy of network ensembles. *Physical Review E*, 79(3):036114.
[33] Bianconi, G. (2013). Statistical mechanics of multiplex networks: Entropy and overlap. *Physical Review E*, 87(6):62806.
[34] Bianconi, G., Darst, R. K., Iacovacci, J., and Fortunato, S. (2014). Triadic closure as a basic generating mechanism of communities in complex networks. *Physical Review E*, 90(4):42806.
[35] Blondel, V. D., Decuyper, A., and Krings, G. (2015). A survey of results on mobile phone datasets analysis. *EPJ Data Science*, 4(1):10.
[36] Blonder, B., and Dornhaus, A. (2011). Time-ordered networks reveal limitations to information flow in ant colonies. *PLOS ONE*, 6(5):e20298.
[37] Boccaletti, S., Bianconi, G., Criado, R., del Genio, C., Gómez-Gardeñes, J., Romance, M., Sendiña-Nadal, I., Wang, Z., and Zanin, M. (2014). The structure and dynamics of multilayer networks. *Physics Reports*, 544(1):1–122.
[38] Boccaletti, S., Latora, V., Moreno, Y., and Chavez, M. (2006). Complex networks: Structure and dynamics. *Physics Reports*, 424(4):175–308.

[39] Boguñá, M., Pastor-Satorras, R., Díaz-Guilera, A., and Arenas, A. (2004). Models of social networks based on social distance attachment. *Physics Review E*, 70(5):56122.

[40] Bollobás, B. (1990). Almost every graph has reconstruction number three. *Journal of Graph Theory*, 14(1):1–4.

[41] Bollobás, B. (2006). *The Art of Mathematics: Coffee Time in Memphis*. Cambridge University Press, Cambridge.

[42] Bonacich, P. (1972). Factoring and weighting approaches to status scores and clique identification. *Journal of Mathematical Sociology*, 2(1):113–20.

[43] Bonacich, P. (1972). Technique for analyzing overlapping memberships. *Sociological Methodology*, 4:176–85.

[44] Bordenave, C., Feige, U., and Mossel, E. (2016). Shotgun assembly of random jigsaw puzzles. arXiv:1605.03086.

[45] Brandes, U. (2001). A faster algorithm for betweenness centrality. *Journal of Mathematical Sociology*, 25(2):163–77.

[46] Brin, S., and Page, L. (1998). The anatomy of a large-scale hypertextual web search engine. *Computer Networks and ISDN Systems*, 30(1):107–17.

[47] Bródka, P., and Kazienko, P. (2014). *Multi-Layered Social Networks*. Springer, Berlin.

[48] Buldyrev, S. V., Parshani, R., Paul, G., Stanley, H. E., and Havlin, S. (2010). Catastrophic cascade of failures in interdependent networks. *Nature*, 464(7291):1025– 8.

[49] Burda, Z., Duda, J., Luck, J. M., and Waclaw, B. (2009). Localization of the maximal entropy random walk. *Physical Review Letters*, 102(16):160602.

[50] Burda, Z., Krzywicki, A., Martin, O. C., and Zagorski, M. (2011). Motifs emerge from function in model gene regulatory networks. *Proceedings of the National Academy of Sciences of the United States of America*, 108(42):17263–8.

[51] Bureau van Dijk. (2010). *Orbis*. Available at https://www.bvdinfo.com/en-us/our-products/company-information/international-products/orbis (accessed May 31, 2018).

[52] Bustos, S., Gomez, C., Hausmann, R., and Hidalgo, C. A. (2012). The dynamics of nestedness predicts the evolution of industrial ecosystems. *PLOS ONE*, 7(11):e49393.

[53] Cairncross, F., and Cairncross, F. C. (2001). *The Death of Distance: How the Communications Revolution Is Changing our Lives*. Harvard Business School, Boston.

[54] Caldarelli, G. (2007). *Scale-Free Networks: Complex Webs in Nature and Technology*. Oxford University Press, Oxford.

[55] Callaghan, T., Mucha, P. J., and Porter, M. A. (2007). Random walker ranking for NCAA Division I-A football. *American Mathematical Monthly*, 114(9):761–77.

[56] Cardillo, A., Gómez-Gardeñes, J., Zanin, M., Romance, M., Papo, D., Del Pozo, F., and Boccaletti, S. (2013). Emergence of network features from multiplexity. *Scientific Reports*, 3:1344.

[57] Cattuto, C., Schmitz, C., Baldassarri, A., Servedio, V. D. P., Loreto, V., Hotho, A., Grahl, M., and Stumme, G. (2007). Network properties of folksonomies. *AI Communications*, 20(4):245–62.

[58] Chatterjee, S., and Diaconis, P. (2013). Estimating and understanding exponential random graph models. *Annals of Statistics*, 41(5):2428–61.

[59] Chatterjee, S., Diaconis, P., and Sly, A. (2011). Random graphs with a given degree sequence. *Annals of Applied Probability*, 21(4):1400–35.

[60] Christen, P. (2012). *Data Matching: Concepts and Techniques for Record Linkage, Entity Resolution, and Duplicate Detection*. Data-Centric Systems and Applications. Springer-Verlag, Berlin, Heidelberg.

[61] Chung, F. R. K. (1999). Review of spectral graph theory. *ACM SIGACT News*, 30(2):14–16.
[62] Cohen, A. M., Hersh, W. R., Dubay, C., and Spackman, K. (2005). Using co-occurrence network structure to extract synonymous gene and protein names from MEDLINE abstracts. *BMC Bioinformatics*, 6(1):103.
[63] Colomer-de Simón, P., Serrano, M. Á., Beiró, M. G., Alvarez-Hamelin, J. I., and Boguñá, M. (2013). Deciphering the global organization of clustering in real complex networks. *Scientific Reports*, 3:2517.
[64] Content, J. and Frenken, K. (2016). Related variety and economic development: A literature review. *European Planning Studies*, 24(12):2097–12.
[65] Costanzo, M., Baryshnikova, A., Bellay, J., Kim, Y., Spear, E. D., Sevier, C. S., Ding, H., Koh, J. L. Y., Toufighi, K., Mostafavi, S., et al. (2010). The genetic landscape of a cell. *Science*, 327(5964):425–31.
[66] Cozzo, E., Kivelä, M., De Domenico, M., Solé-Ribalta, A., Arenas, A., Gómez, S., Porter, M. A., and Moreno, Y. (2015). Structure of triadic relations in multiplex networks. *New Journal of Physics*, 17(7):73029.
[67] Czabarka, É., Dutle, A., Erdős, P. L., and Miklós, I. (2015). On realizations of a joint degree matrix. *Discrete Applied Mathematics*, 181:283–8.
[68] da Fontoura Costa, L., and Travieso, G. (2007). Exploring complex networks through random walks. *Physical Review E*, 75(1):16102.
[69] Daqing, L., Kosmidis, K., Bunde, A., and Havlin, S. (2011). Dimension of spatially embedded networks. *Nature Physics*, 7(6):481–4.
[70] Darst, R. K., Granell, C., Arenas, A., Gómez, S., Saramäki, J., and Fortunato, S. (2016). Detection of timescales in evolving complex systems. arXiv:1604.00758.
[71] De Domenico, M., Porter, M. A., and Arenas, A. (2015). MuxViz: A tool for multilayer analysis and visualization of networks. *Journal of Complex Networks*, 3(2):159–76.
[72] De Domenico, M., Solé-Ribalta, A., Cozzo, E., Kivelä, M., Moreno, Y., Porter, A., Gómez, S., and Arenas, A. (2013). Mathematical formulation of multilayer networks. *Physical Review X*, 3(4):41022.
[73] De Domenico, M., Solé-Ribalta, A., Gómez, S., and Arenas, A. (2014). Navigability of interconnected networks under random failures. *Proceedings of the National Academy of Sciences of the United States of America*, 111(23):8351–6.
[74] De Domenico, M., Solé-Ribalta, A., Omodei, E., Gómez, S., and Arenas, A. (2015). Ranking in interconnected multilayer networks reveals versatile nodes. *Nature Communications*, 6:6868.
[75] de Simon, P. C. (2014). RandNetGen: A random network generator. Available at http://github.com/polcolomer/RandNetGen (accessed May 31, 2018).
[76] Del Genio, C. I., Kim, H., Toroczkai, Z., and Bassler, K. E. (2010). Efficient and exact sampling of simple graphs with given arbitrary degree sequence. *PLOS ONE*, 5(4):e10012.
[77] Dickison, M., Havlin, S., and Stanley, H. E. (2012). Epidemics on interconnected networks. *Physical Review E*, 85(6):66109.
[78] Diestel, R. (2000). *Graph Theory*. Springer-Verlag, Berlin, Heidelberg.
[79] Dimitropoulos, X., Krioukov, D., Riley, G., and Vahdat, A. (2009). Graph annotations in modeling complex network topologies. *ACM Transactions on Modeling and Computer Simulation*, 19(4):17.
[80] Dorogovtsev, S. N., Mendes, J., and Samukhin, A. (2001). Size-dependent degree distribution of a scale-free growing network. *Physics Review E*, 63(6):62101.
[81] Dunbar, R. I. M. (2011). Constraints on the evolution of social institutions and their implications for information flow. *Journal of Institutional Economics*, 7(Special Issue 3): 345–71.

[82] Eagle, N., Pentland, A. S., and Lazer, D. (2009). Inferring friendship network structure by using mobile phone data. *Proceedings of the National Academy of Sciences of the United States of America*, 106(36):15274–8.

[83] Eckmann, J.-P., Moses, E., and Sergi, D. (2004). Entropy of dialogues creates coherent structures in e-mail traffic. *Proceedings of the National Academy of Sciences of the United States of America*, 101(40):14333–7.

[84] Edmonds, P. (1997). Choosing the word most typical in context using a lexical co-occurrence network. In *Proceedings of the Eighth Conference on European Chapter for Computational Linguistics*, pp. 507–9. Association for Computational Linguistics, Stroudsburg.

[85] Eguíluz, V., Chialvo, D., Cecchi, G., Baliki, M., and Apkarian, A. V. (2005). Scale-free brain functional networks. *Physics Review Letters*, 94(1):18102.

[86] Ellis, D., Friedgut, E., Kindler, G., and Yehudayoff, A. (2016). Geometric stability via information theory. *Discrete Analysis*, 2016:10.

[87] ESPON (2010). *FOCI: Future Orientation for Cities (Report).* Available at http://www.espon.eu/main/ (accessed May 31, 2018).

[88] Estrada, E. (2011). *The Structure of Complex Networks: Theory and Applications*. Oxford University Press, Oxford.

[89] Eurostat. (2008). *NACE Rev 2: Statistical Classification of Economic Activities in the European Community.* Available at http://ec.europa.eu/eurostat/documents/3859598/5902521/KS-RA-07-015-EN.PDF (accessed May 31, 2018).

[90] Expert, P., Evans, T. S., Blondel, V. D., and Lambiotte, R. (2011). Uncovering space-independent communities in spatial networks. *Proceedings of the National Academy of Sciences of the United States of America*, 108(19):7663–8.

[91] Fagerland, M. W. (2012). *t*-Tests, non-parametric tests, and large studies: A paradox of statistical practice? *BMC Medical Research Methodology*, 12(78):1–7.

[92] Feldman, R., and Sanger, J. (2006). *Text Mining Handbook: Advanced Approaches in Analyzing Unstructured Data.* Cambridge University Press, New York.

[93] Ferrer i Cancho, R., and Ricard, V. S. (2001). The small world of human language. *Proceedings of the Royal Society B*, 268(1482):2261–5.

[94] Fienberg, S. E., Meyer, M. M., and Wasserman, S. S. (1985). Statistical analysis of multiple sociometric relations. *Journal of the American Statistical Association*, 80(389):51–67.

[95] Flaounas, I., Ali, O., Turchi, M., Snowsill, T., Nicart, F., De Bie, T., and Cristianini, (2011). NOAM: News outlets analysis and monitoring system. In *Proceedings of the 2011 ACM SIGMOD International Conference on Management of Data*, pp. 1275–8. ACM, New York.

[96] Fortunato, S. (2010). Community detection in graphs. *Physics Reports*, 486(3–5):75–174.

[97] Foster, D., Foster, J., Paczuski, M., and Grassberger, P. (2010). Communities, clustering phase transitions, and hysteresis: Pitfalls in constructing network ensembles. *Physics Review E*, 81(4):46115.

[98] Foster, D. V., Foster, J. G., Grassberger, P., and Paczuski, M. (2011). Clustering drives assortativity and community structure in ensembles of networks. *Physics Review E*, 84(6):66117.

[99] Fournet, J., and Barrat, A. (2014). Contact patterns among high school students. *PLOS ONE*, 9(9):e107878.

[100] Freeman, L. C. (1977). A set of measures of centrality based on betweenness. *Sociometry*, 40(1):35–41.

[101] Freeman, L. C. (1979). Centrality in social networks conceptual clarification. *Social Networks*, 1(3):215–39.

[102] Freilich, S., Kreimer, A., Meilijson, I., Gophna, U., Sharan, R., and Ruppin, E. (2010). The large-scale organization of the bacterial network of ecological co- occurrence interactions. *Nucleic Acids Research*, 38(12):3857–68.

[103] Frenken, K., Van Oort, F., and Verburg, T. (2007). Related variety, unrelated variety and regional economic growth. *Regional Studies*, 41(5):685–97.
[104] Gallos, L. K., and Argyrakis, P. (2003). Distribution of infected mass in disease spreading in scale-free networks. *Physica A*, 330(1):117–23.
[105] Gallotti, R., Porter, M. A., and Barthélemy, M. (2016). Lost in transportation: Information measures and cognitive limits in multilayer navigation. *Science Advances*, 2(2):e1500445.
[106] Gao, J., Buldyrev, S. V., Stanley, H. E., and Havlin, S. (2011). Networks formed from interdependent networks. *Nature Physics*, 8(1):40–8.
[107] Garas, A. (2016). *Interconnected Networks*. Springer, Cham.
[108] Garas, A., Argyrakis, P., Rozenblat, C., Tomassini, M., and Havlin, S. (2010). Worldwide spreading of economic crisis. *New Journal of Physics*, 12(11):113043.
[109] Gauvin, L., Panisson, A., Barrat, A., and Cattuto, C. (2015). Revealing latent factors of temporal networks for mesoscale intervention in epidemic spread. arXiv:1501.02758.
[110] Gauvin, L., Panisson, A., and Cattuto, C. (2014). Detecting the community structure and activity patterns of temporal networks: A non-negative tensor factorization approach. *PLOS ONE*, 9(1):e86028.
[111] Gauvin, L., Panisson, A., Cattuto, C., and Barrat, A. (2013). Activity clocks: Spreading dynamics on temporal networks of human contact. *Scientific Reports*, 3:3099.
[112] Gemmetto, V., Barrat, A., and Cattuto, C. (2014). Mitigation of infectious disease at school: Targeted class closure vs school closure. *BMC Infectious Diseases*, 14(1):1–10.
[113] Génois, M., Vestergaard, C. L., Cattuto, C., and Barrat, A. (2015). Compensating for population sampling in simulations of epidemic spread on temporal contact networks. *Nature Communications*, 6:8860.
[114] Génois, M., Vestergaard, C. L., Fournet, J., Panisson, A., Bonmarin, I., and Barrat, A. (2015). Data on face-to-face contacts in an office building suggest a low-cost vaccination strategy based on community linkers. *Network Science*, 3(03):326–47.
[115] Gfeller, D., and De Los Rios, P. (2007). Spectral coarse graining of complex networks. *Physical Review Letters*, 99(3):38701.
[116] Ghoshal, G., Zlatíc, V., Caldarelli, G., and Newman, M. E. J. (2009). Random hypergraphs and their applications. *Physical Review E*, 79(6):66118.
[117] Gjoka, M., Kurant, M., and Markopoulou, A. (2013). 2.5K-graphs: From sampling to generation. In *INFOCOM, 2013 Proceedings IEEE*, pp. 1968–76. IEEE, Piscataway.
[118] Goldenberg, J., and Levy, M. (2009). Distance is not dead: Social interaction and geographical distance in the internet era. arXiv:0906.3202.
[119] Gómez, S., Díaz-Guilera, A., Gómez-Gardeñes, J., Pérez-Vicente, C. J., Moreno, Y., and Arenas, A. (2013). Diffusion dynamics on multiplex networks. *Physical Review Letters*, 110(2):028701.
[120] Gómez-Gardeñes, J., Zamora-López, G., Moreno, Y., and Arenas, A. (2010). From modular to centralized organization of synchronization in functional areas of the cat cerebral cortex. *PLOS ONE*, 5(8):e12313.
[121] González, M. C., Hidalgo, C. A., and Barabási, A.-L. (2008). Understanding individual human mobility patterns. *Nature*, 453(7196):779–82.
[122] Granell, C., Gómez, S., and Arenas, A. (2013). Dynamical interplay between awareness and epidemic spreading in multiplex networks. *Physical Review Letters*, 111(12):128701.
[123] Granovetter, M. S. (1973). The strength of weak ties. *American Journal of Sociology*, 78(6):1360–80.
[124] Guimerà, R., Sales-Pardo, M., and Amaral, L. A. (2007). Classes of complex networks defined by role-to-role connectivity profiles. *Nature Physics*, 3(1):63–9.

[125] Halu, A., Mondragón, R. J., Panzarasa, P., and Bianconi, G. (2013). Multiplex PageRank. *PLOS ONE*, 8(10):e78293.

[126] Harary, F. (1964). On the reconstruction of a graph from a collection of subgraphs. In *Theory of Graphs and its Applications: Proceedings of the Symposium Held in Smolenice in June 1963)*, pp. 47–52. Publishing House of the Czechoslovak Academy of Sciences, Prague.

[127] Herfindahl, O. C. (1950). Concentration in the US steel industry. PhD thesis, Columbia University, New York.

[128] Hicks, J., Traag, V. A., and Reinanda, R. (2015). Turning digitised newspapers into networks of political elites. *Asian Journal of Social Science*, 43(5):567–87.

[129] Hidalgo, C. A., and Hausmann, R. (2009). The building blocks of economic complexity. *Proceedings of the National Academy of Sciences of the United States of America*, 106(26):10570–5.

[130] Hirschman, A. O. (1980). *National Power and the Structure of Foreign Trade*, volume 105. University of California Press, Berkeley.

[131] Hizanidis, J., Kouvaris, N. E., Gorka, Z.-L., Díaz-Guilera, A., and Antonopoulos, C. G. (2016). Chimera-like states in modular neural networks. *Scientific Reports*, 6: 19845.

[132] Hoover, E. M. (1948). *The Location of Economic Activity*. McGraw-Hill Book Company, Inc., London.

[133] Horvát, S., Czabarka, É., and Toroczkai, Z. (2015). Reducing degeneracy in maximum entropy models of networks. *Physics Review Letters*, 114(15–17):158701.

[134] Isard, W. (1956). *Localization and Space Economy: A General Theory Relating to Industrial Location, Market Areas, Land Use, Trade And Urban Structure*. MIT Press, Cambridge, MA.

[135] Isella, L., Stehlé, J., Barrat, A., Cattuto, C., Pinton, J.-F., and den Broeck, W. (2011). What's in a crowd? Analysis of face-to-face behavioral networks. *Journal of Theoretical Biology*, 271(1):166–80.

[136] Jackson, M. O. (2010). *Social and Economic Networks*. Princeton University Press, Princeton.

[137] Jamakovic, A., Mahadevan, P., Vahdat, A., Boguñá, M., and Krioukov, D. (2009). How small are building blocks of complex networks. arXiv:0908.1143.

[138] Jo, H.-H., Baek, S. K., and Moon, H.-T. (2006). Immunization dynamics on a two-layer network model. *Physica A*, 361(2):534–42.

[139] Jo, H.-H., Karsai, M., Karikoski, J., and Kaski, K. (2012). Spatiotemporal correlations of handset-based service usages. *EPJ Data Science*, 1(1):10.

[140] Jo, H.-H., Pan, R. K., Perotti, J. I., and Kaski, K. (2013). Contextual analysis framework for bursty dynamics. *Physical Review E*, 87:062131.

[141] Jo, H.-H., Saramäki, J., Dunbar, R. I. M., and Kaski, K. (2014). Spatial patterns of close relationships across the lifespan. *Scientific Reports*, 4:6988.

[142] Juršič, M., Sluban, B., Cestnik, B., Grčar, M., and Lavrač, N. (2012). Bridging concept identification for constructing information networks from text documents. In Berthold M. R. (ed.), *Bisociative Knowledge Discovery*, pp. 66–90. Springer, Berlin, Heidelberg.

[143] Karsai, M., Kivelä, M., Pan, R. K., Kaski, K., Kertész, J., Barabási, A.-L., and Saramäki, J. (2011). Small but slow world: How network topology and burstiness slow down spreading. *Physical Review E*, 83:025102.

[144] Katz, L. (1953). A new status index derived from sociometric analysis. *Psychometrika*, 18(1):39–43.

[145] Kelly, P. J. (1957). A congruence theorem for trees. *Pacific Journal of Mathematics*, 7(1):961–8.

[146] Kermack, W. O., and McKendrick, A. G. (1927). A contribution to the mathematical theory of epidemics. *Proceedings of the Royal Society A*, 115(772):700–21.

[147] Kim, H., Del Genio, C. I., Bassler, K. E., and Toroczkai, Z. (2012). Constructing and sampling directed graphs with given degree sequences. *New Journal of Physics*, 14(2): 23012.

[148] Kim, H., Toroczkai, Z., Erdős, P. L., Miklós, I., and Székely, L. A. (2009). Degree-based graph construction. *Journal of Physics A*, 42(39):392001.

[149] Kivelä, M., Arenas, A., Barthélemy, M., Gleeson, J. P., Moreno, Y., and Porter, M. A. (2014). Multilayer networks. *Journal of Complex Networks*, 2(3):203–71.

[150] Kleinberg, J. M. (1999). Authoritative sources in a hyperlinked environment. *Journal of the ACM*, 46(5):604–32.

[151] Klemm, K., and Eguíluz, V. (2002). Highly clustered scale-free networks. *Physics Review E*, 65(3):36123.

[152] Klimek, P., Diakonova, M., Eguiluz, V., Miguel, M. S., and Thurner, S. (2016). Dynamical origins of the community structure of multi-layer societies. *New Journal of Physics*, 18(8):1–8.

[153] Kok, S., and Domingos, P. M. (2008). Extracting semantic networks from text via relational clustering. In Daelemans, W., Goethals, B., and Morik, K. (eds), *Machine Learning and Knowledge Discovery in Databases*, pp. 624–39. Springer, Berlin, Heidelberg.

[154] Kolda, T. G., and Bader, B. W. (2009). Tensor decompositions and applications. *SIAM Review*, 51(3):455–500.

[155] Kosmidis, K., Havlin, S., and Bunde, A. (2008). Structural properties of spatially embedded networks. *Europhysics Letters*, 82(4):48005.

[156] Kossinets, G., and Watts, D. J. (2006). Empirical analysis of an evolving social network. *Science*, 311(5757):88–90.

[157] Kouvaris, N. E., Hata, S., and Díaz-Guilera, A. (2015). Pattern formation in multiplex networks. *Scientific Reports*, 5:10840.

[158] Kralj Novak, P., Grčar, M., and Mozetič, I. (2015). *Analysis of Financial News with NewsStream*. Technical Report IJS-DP-11892. Jožef Stefan Institute, Ljubljana.

[159] Krings, G., Calabrese, F., Ratti, C., and Blondel, V. D. (2009). Urban gravity: A model for inter-city telecommunication flows. *Journal of Statistical Mechanics*, 2009(7):L07003.

[160] Krugman, P. (1993). First nature, second nature, and metropolitan location. *Journal of Regional Science*, 33(2):129–44.

[161] Krzywinski, M., Birol, I., Jones, S. J. M., and Marra, M. A. (2012). Hive plots: Rational approach to visualizing networks. *Briefings in Bioinformatics*, 13(5):627–44.

[162] Kumpula, J. M., Onnela, J. P., Saramäki, J., Kaski, K., and Kertész, J. (2007). Emergence of communities in weighted networks. *Physical Review Letters*, 99(22):228701.

[163] Kuperman, M., and Abramson, G. (2001). Small world effect in an epidemiological model. *Physical Review Letters*, 86(13):2909.

[164] Kuramoto, Y., and Battogtokh, D. (2002). Coexistence of coherence and incoherence in nonlocally coupled phase oscillators. *Nonlinear Phenomena in Complex Systems*, 5(4):380–5.

[165] Kwak, H., Lee, C., Park, H., and Moon, S. (2010). What is Twitter, a social network or a news media? In *Proceedings of the 19th International World Wide Web Conference*, pp. 591–600. ACM, New York.

[166] Lambiotte, R., Blondel, V., Dekerchove, C., Huens, E., Prieur, C., Smoreda, Z., and Vandooren, P. (2008). Geographical dispersal of mobile communication networks. *Physica A*, 387(21):5317–25.

[167] Lambiotte, R., Delvenne, J.-C., and Barahona, M. (2014). Random walks, Markov processes and the multiscale modular organization of complex networks. *IEEE Transactions on Network Science and Engineering*, 1(2):76–90.

[168] Lazer, D., Pentland, A., Adamic, L. A., Aral, S., Barabasi, A.-L., Brewer, D., Christakis, N., Contractor, N., Fowler, J., Gutmann, M., et al. (2009). Computational social science. *Science*, 323(5915):721–3.

[169] Leban, G., Fortuna, B., Brank, J., and Grobelnik, M. (2014). Event registry: Learning about world events from news. In *Proceedings of the 23rd International World Wide Web Conference*, pp. 107–10. ACM, New York.

[170] Lee, K.-M., Kim, J. Y., Cho, W.-K., Goh, K.-I., and Kim, I. M. (2012). Correlated multiplexity and connectivity of multiplex random networks. *New Journal of Physics*, 14(3):33027.

[171] Lee, S. H., Kim, P.-J., and Jeong, H. (2006). Statistical properties of sampled networks. *Physics Review E*, 73(1):016102.

[172] Leetaru, K. H. (2011). Culturomics 2.0: Forecasting large-scale human behavior using global news media tone in time and space. *First Monday*, 16(9): http://www.firstmonday.dk/ojs/index.php/fm/article/view/3663/3040.

[173] Lengyel, B., Varga, A., Sagvari, B., Jakobi, A., and Kertesz, J. (2015). Geographies of an online social network. *PLOS ONE*, 10(9): e0137248.

[174] Lewis, K., Kaufman, J., Gonzalez, M., Wimmer, A., and Christakis, N. (2008). Tastes, ties, and time: A new social network dataset using Facebook.com. *Social Networks*, 30(4):330–42.

[175] Li, W., Bashan, A., Buldyrev, S. V., Stanley, H. E., and Havlin, S. (2012). Cascading failures in interdependent lattice networks: The critical role of the length of dependency links. *Physics Review Letters*, 108(22):228702.

[176] Lima, A., De Domenico, M., Pejovic, V., and Musolesi, M. (2013). Exploiting cellular data for disease containment and information campaigns strategies in country-wide epidemics. arXiv:1306.4534.

[177] Liu, B. (2015). *Sentiment Analysis: Mining Opinions, Sentiments, and Emotions*. Cambridge University Press, Cambridge.

[178] Liu, H., and Cong, J. (2013). Language clustering with word co-occurrence networks based on parallel texts. *Chinese Science Bulletin*, 58(10):1139–44.

[179] Lloyd, L., Kechagias, D., and Skiena, S. (2005). Lydia: A system for large-scale news analysis. In Consens M., and Navarro G. (eds), *String Processing and Information Retrieval*, pp. 161–6. Springer, Berlin, Heidelberg.

[180] Lovász, L. (1993). Random walks on graphs: A survey. *Combinatorics, Paul Erdos is Eighty*, 2(1):1–46.

[181] Machens, A., Gesualdo, F., Rizzo, C., Tozzi, A. E., Barrat, A., and Cattuto, C. (2013). An infectious disease model on empirical networks of human contact: Bridging the gap between dynamic network data and contact matrices. *BMC Infectious Diseases*, 13(1):185.

[182] Mahadevan, P., Krioukov, D., Fall, K., and Vahdat, A. (2006). Systematic topology analysis and generation using degree correlations. *ACM SIGCOMM Computer Communication Review*, 36(4):135–46.

[183] Mahadevan, P., Krioukov, D., Fomenkov, M., Huffaker, B., Dimitropoulos, X., Claffy, K., and Vahdat, A. (2006b). The Internet AS-level topology: Three data sources and one definitive metric. *ACM SIGCOMM Computer Communication Review*, 36(1):17–26.

[184] Mane, K. K., and Börner, K. (2004). Mapping topics and topic bursts in PNAS. *Proceedings of the National Academy of Sciences of the United States of America*, 101(Suppl. 1):5287–90.

[185] Martinsson, A. (2017). A linear threshold for uniqueness of solutions to random jigsaw puzzles. arXiv:1701.04813.

[186] Maslov, S., Sneppen, K., and Alon, U. (2003). Correlation profiles and motifs in complex networks. In Bornholdt, S., and Schuster, H. G. (eds), Handbook of Graphs and Networks: From the Genome to the Internet. Wiley-VCH, Berlin.

[187] Maslov, S., Sneppen, K., and Zaliznyak, A. (2004). Detection of topological patterns in complex networks: Correlation profile of the internet. *Physica A*, 333(1–4):529–40.
[188] Mastrandrea, R., and Barrat, A. (2016). How to estimate epidemic risk from incomplete contact diaries data? *PLoS Computational Biology*, 12(6):1–19.
[189] Mastrandrea, R., Fournet, J., and Barrat, A. (2015). Contact patterns in a high school: A comparison between data collected using wearable sensors, contact diaries and friendship surveys. *PLOS ONE*, 10(9):1–26.
[190] McPherson, M., Smith-Lovin, L., and Cook, J. M. (2001). Birds of a feather: Homophily in social networks. *Annual Review of Sociology*, 27(1):415–44.
[191] Mercer. (2012). Quality of Living City Ranking. Mercer, New York.
[192] Miljkovíc, D., Stare, T., Mozetič, I., Podpečan, V., Petek, M., Witek, K., Dermastia, M., Lavrač, N., and Gruden, K. (2012). Signalling network construction for modelling plant defence response. *PLOS ONE*, 7(12):e0051822.
[193] Miller, G. A. (1995). WordNet: A lexical database for English. *Communications of the ACM*, 38(11):39–41.
[194] Miller, M., Sathi, C., Wiesenthal, D., Leskovec, J., and Potts, C. (2011). Sentiment flow through hyperlink networks. In *Proceedings of the Fifth International Conference on Weblogs and Social Media*, pp. 550–3. AAAI Press, Palo Alto.
[195] Milo, R., Shen-Orr, S., Itzkovitz, S., Kashtan, N., Chklovskii, D., and Alon, U. (2002). Network motifs: Simple building blocks of complex networks. *Science*, 298(5594):824–7.
[196] OECD/Eurostat. (2005). *Oslo Manual: Guidelines for Collecting and Interpreting Innovation Data*. OECD Publishing, Paris.
[197] Mossel, E., and Ross, N. (2015). Shotgun assembly of labeled graphs. arXiv:1504.07682.
[198] Motter, A. E., and Toroczkai, Z. (2007). Introduction: Optimization in networks. *Chaos*, 17(2):26101.
[199] Mozetič, I., Grčar, M., and Smailovíc, J. (2016). Multilingual Twitter sentiment classification: The role of human annotators. *PLOS ONE*, 11(5):e0155036.
[200] Mucha, P. J., Richardson, T., Macon, K., Porter, M. A., and Onnela, J.-P. (2010). Community structure in time-dependent, multiscale, and multiplex networks. *Science*, 328(5980):876–8.
[201] Murase, Y., Jo, H.-H., Török, J., Kertész, J., and Kaski, K. (2015). Modeling the role of relationship fading and breakup in social network formation. *PLOS ONE*, 10(7):e0133005.
[202] Murase, Y., Török, J., Jo, H.-H., Kaski, K., and Kertész, J. (2014). Multilayer weighted social network model. *Physical Review E*, 90(5):052810.
[203] Murase, Y., Uchitane, T., and Ito, N. (2014). A tool for parameter-space explorations. *Physics Procedia*, 57:73–6.
[204] Myers, S. A., Sharma, A., Gupta, P., and Lin, J. (2014). Information network or social network? The structure of the Twitter follow graph. In *Proceedings of the 23rd International World Wide Web Conference*, pp. 493–8. ACM, New York.
[205] Nakao, H., and Mikhailov, A. S. (2010). Turing patterns in network-organized activator-inhibitor systems. *Nature Physics*, 6(7):544–50.
[206] Nelsen, R. B. (2007). *An Introduction to Copulas*. Springer, New York.
[207] Nenadov, R., Pfister, P., and Steger, A. (2016). Unique reconstruction threshold for random jigsaw puzzles. arXiv:1605.03043.
[208] Newman, M. E. (2002). Assortative mixing in networks. *Physical Review Letters*, 89(20):1–5.
[209] Newman, M. E. (2003). The structure and function of complex networks. *SIAM Review*, 45(2):167–256.
[210] Newman, M. E. (2010). *Networks: An Introduction*. Oxford University Press, New York.

[211] Newman, M. E. J. (2005). A measure of betweenness centrality based on random walks. *Social Networks*, 27(1):39–54.

[212] Newman, M. E. J. (2009). Random graphs with clustering. *Physics Review Letters*, 103(5): 1–4.

[213] Nicosia, V., Bianconi, G., Latora, V., and Barthelemy, M. (2013). Growing multiplex networks. *Physical Review Letters*, 111(5):58701.

[214] Noh, J. D., and Rieger, H. (2004). Random walks on complex networks. *Physical Review Letters*, 92(11):118701.

[215] Ohlin, B. G. (1933). *Interregional and International Trade.* Harvard University Press, Cambridge, MA.

[216] Onnela, J.-P., Arbesman, S., González, M. C., Barabási, A.-L., and Christakis, N. A. (2011). Geographic constraints on social network groups. *PLOS ONE*, 6(4):e16939.

[217] Onnela, J.-P., Saramäki, J., Hyvönen, J., Szabó, G., de Menezes, M., Kaski, K., Barabási, A.-L., and Kertész, J. (2007). Analysis of a large-scale weighted network of one-to-one human communication. *New Journal of Physics*, 9(6):179.

[218] Onnela, J.-P., Saramäki, J., Hyvönen, J., Szabó, G., Lazer, D., Kaski, K., Kertész, J., and Barabási, A.-L. (2007). Structure and tie strengths in mobile communication networks. *Proceedings of the National Academy of Sciences of the United States of America*, 104(18): 7332–6.

[219] Orsini, C., Dankulov, M. M., Colomer-de Simón, P., Jamakovic, A., Mahadevan, P., Vahdat, A., Bassler, K. E., Toroczkai, Z., Boguñá, M., Caldarelli, G., et al. (2015). Quantifying randomness in real networks. *Nature Communications*, 6:8627.

[220] Özgür, A., Cetin, B., and Bingol, H. (2008). Co-occurrence network of Reuters news. *International Journal of Modern Physics C*, 19(05):689–702.

[221] Padgett, J. F., and Ansell, C. K. (1993). Robust action and the rise of the Medici, 1400–1434. *American Journal of Sociology*, 98(6):1259–319.

[222] Palchykov, V., Kaski, K., Kertész, J., Barabási, A.-L., and Dunbar, R. I. M. (2012). Sex differences in intimate relationships. *Scientific Reports*, 2:370.

[223] Pan, J., and Singleton, K. J. (2008). Default and recovery implicit in the term structure of sovereign CDS spreads. *The Journal of Finance*, 63(5):2345–84.

[224] Panaggio, M. J., and Abrams, D. M. (2015). Chimera states: Coexistence of coherence and incoherence in networks of coupled oscillators. *Nonlinearity*, 28(3):R67.

[225] Papadopoulos, F., Kitsak, M., Serrano, M. Á., Boguñá, M., and Krioukov, D. (2012). Popularity versus similarity in growing networks. *Nature*, 489(7417):537–40.

[226] Pearson, K. (1895). Note on regression and inheritance in the case of two parents. *Proceedings of the Royal Society of London*, 58(347–52):240–2.

[227] Pebody, L. (2004). The reconstructibility of finite abelian groups. *Combinatorics, Probability and Computing*, 13(6):867–92.

[228] Pebody, L. (2007). Reconstructing odd necklaces. *Combinatorics, Probability and Computing*, 16(4):503–14.

[229] Pebody, L., Radcliffe, A. J., and Scott, A. D. (2003). Finite subsets of the plane are 18-reconstructible. *SIAM Journal on Discrete Mathematics*, 16(2):272–5.

[230] Pfitzner, R., Scholtes, I., Garas, A., Tessone, C. J., and Schweitzer, F. (2013). Betweenness preference: Quantifying correlations in the topological dynamics of temporal networks. *Physical Review Letters*, 110(19):198701.

[231] Piškorec, M., Sluban, B., and Šmuc, T. (2015). MultiNets: Web-based multilayer network visualization. In Bifet, A., May, M., Zadrozny, B., Gavalda, R., Pedreschi, D., Bonchi, F.,

Cardoso, J., and Spiliopoulou, M. (eds), *Machine Learning and Knowledge Discovery in Databases*, pp. 298–302. Springer, Cham.

[232] Popović, M., Štefančić, H., Sluban, B., Novak, P. K., Grčar, M., Mozetič, I., Puliga, M., and Zlatić, V. (2014). Extraction of temporal networks from term co-occurrences in online textual sources. *PLOS ONE*, 9(12):e99515.

[233] United Nations Global Pulse. (2013). *Mobile Phone Network Data for Development*. Available at http://www.slideshare.net/unglobalpulse/mobile-data-for-development-primer-october-2013 (accessed May 31, 2017).

[234] Pumain, D., Paulus, F., Vacchiani-Marcuzzo, C., and Lobo, J. (2006). An evolutionary theory for interpreting urban scaling laws. *Cybergeo*, Systèmes, Modélisation, Géostatistiques (343): http://journals.openedition.org/cybergeo/2519.

[235] Radcliffe, A. J., and Scott, A. D. (1998). Reconstructing subsets of $Z_n$. *Journal of Combinatorial Theory, Series A*, 83(2):169–87.

[236] Radicchi, F., and Arenas, A. (2013). Abrupt transition in the structural formation of interconnected networks. *Nature Physics*, 9(11):717–20.

[237] Raghavan, V., Bollmann, P., and Jung, G. S. (1989). A critical investigation of recall and precision as measures of retrieval system performance. *ACM Transactions on Information Systems*, 7(3):205–29.

[238] Ranco, G., Aleksovski, D., Caldarelli, G., Grčar, M., and Mozetič, I. (2015). The effects of Twitter sentiment on stock price returns. *PLOS ONE*, 10(9):e0138441.

[239] Roberts, E. S., and Coolen, A. C. C. (2014). Random graph ensembles with many short loops. *ESAIM: Proceedings and Surveys*, 47(2014):97–115.

[240] Rohr, R. P., Saavedra, S., and Bascompte, J. (2014). On the structural stability of mutualistic systems. *Science*, 345(6195):1253497.

[241] Rossi, L., and Magnani, M. (2015). Towards effective visual analytics on multiplex and multilayer networks. *Chaos, Solitons & Fractals*, 72:68–76.

[242] Rosvall, M., and Bergstrom, C. T. (2007). An information-theoretic framework for resolving community structure in complex networks. *Proceedings of the National Academy of Sciences of the United States of America*, 104(18):7327–31.

[243] Rosvall, M., Esquivel, A. V. Lancichinetti, A., West, J. D., and Lambiotte, R. (2014). Memory in network flows and its effects on spreading dynamics and community detection. *Nature Communications*, 5:4630.

[244] Lennert, M., Van Hamme, G., Patris, C., Smętkowski, M., Płoszaj, A., Gorzelak, G., Kozak, M., Olechnicka, A., Wojnar, K., Hryniewicz, J., et al. (eds), *FOCI: Future Orientation for Cities*. ESPON Final report. Available at http://hal.archives-ouvertes.fr/hal-00734406/document (accessed June 4, 2018).

[245] Rozenblat, C. (2015). Inter-cities' multinational firm networks and gravitation model. *Annals of the Association of Economic Geographers*, 61(3):219–37.

[246] Rozenblat, C., Zaidi, F., and Bellwald, A. (2017). The multipolar regionalization of cities in multinational firms' networks. *Global Networks*, 17(2):171–94.

[247] Rozenfeld, H. D., Kirk, J. E., Bollt, E. M., and Ben-Avraham, D. (2005). Statistics of cycles: How loopy is your network? *Journal of Physics A*, 38(21):4589.

[248] Saavedra, S., Stouffer, D. B., Uzzi, B., and Bascompte, J. (2011). Strong contributors to network persistence are the most vulnerable to extinction. *Nature*, 478(7368):233–5.

[249] Sabidussi, G. (1966). The centrality index of a graph. *Psychometrika*, 31(4):581–603.

[250] Saito, K., and Yamada, T. (2006). Extracting communities from complex networks by the *k*-dense method. In Tsumoto, S., Clifton, C. W., Zhong, N., Wu, X., Liu, J., Wah, B. W., and Cheung, Y.-M. (eds), *Sixth IEEE International Conference on Data Mining: Workshops (ICDMW'06)*, pp. 300–4. IEEE, Los Alamitos.

[251] Salton, G. (1989). *Automatic Text Processing: The Transformation, Analysis, and Retrieval of Information by Computer*. Addison-Wesley, Boston.

[252] Samukhin, A. N., Dorogovtsev, S. N., and Mendes, J. F. F. (2008). Laplacian spectra of, and random walks on, complex networks: Are scale-free architectures really important? *Physical Review E*, 77(3):36115.

[253] Sánchez-García, R. J., Cozzo, E., and Moreno, Y. (2014). Dimensionality reduction and spectral properties of multilayer networks. *Physical Review E*, 89(5):52815.

[254] Saramäki, J., Leicht, E. A., López, E., Roberts, S. G. B., Reed-Tsochas, F., and Dunbar, R. I. M. (2014). Persistence of social signatures in human communication. *Proceedings of the National Academy of Sciences of the United States of America*, 111(3):942–7.

[255] Sassen, S. (2001). *The Global City: New York, London, Tokyo*. Princeton University Press, Princeton.

[256] Scholl, T., Garas, A., and Schweitzer, F. (2015). The spatial component of R&D networks. arXiv:1509.08291.

[257] Scholtes, I., Wider, N., and Garas, A. (2015). Higher-order aggregate networks in the analysis of temporal networks: Path structures and centralities. *European Physical Journal B*, 89(3):1–15.

[258] Scholtes, I., Wider, N., Pfitzner, R., Garas, A., Tessone, C. J., and Schweitzer, F. (2014). Causality-driven slow-down and speed-up of diffusion in non-Markovian temporal networks. *Nature Communications*, 5:5024.

[259] Secrier, M., Pavlopoulos, G. A., Aerts, J., and Schneider, R. (2012). Arena3D: Visualizing time-driven phenotypic differences in biological systems. *BMC Bioinformatics*, 13(1):45.

[260] Seidman, S. B. (1983). Network structure and minimum degree. *Social Networks*, 5(3): 269–87.

[261] Seyed-Allaei, H., Bianconi, G., and Marsili, M. (2006). Scale-free networks with an exponent less than two. *Physical Review E*, 73(4):46113.

[262] Shalgi, R., Lieber, D., Oren, M., and Pilpel, Y. (2007). Global and local architecture of the mammalian microRNA–transcription factor regulatory network. *PLoS Computational Biology*, 3(7):e131.

[263] Shanahan, M. (2010). Metastable chimera states in community-structured oscillator networks. *Chaos*, 20(1):13108.

[264] Shang, Y., Li, Y., Lin, H., and Yang, Z. (2011). Enhancing biomedical text summarization using semantic relation extraction. *PLOS ONE*, 6(8):1–10.

[265] Shaw, M. E. (1954). Group structure and the behavior of individuals in small groups. *Journal of Psychology*, 38(1):139–49.

[266] Simpson, E. H. (1949). Measurement of diversity. *Nature*. 163(4148):688.

[267] Sinatra, R., Gómez-Gardeñes, J., Lambiotte, R., Nicosia, V., and Latora, V. (2011). Maximal-entropy random walks in complex networks with limited information. *Physical Review E*, 83(3):30103.

[268] Skiena, S. S. (1998). *The Algorithm Design Manual*. Springer-Verlag New York, New York.

[269] Sluban, B., Grčar, M., and Mozetič, I. (2016). Temporal multi-layer network construction from major news events. In Cherifi H., Gonçalves B., Menezes R., Sinatra R. (eds), *Complex Networks VII: Proceedings of the 7th Workshop on Complex Networks CompleNet*, pp. 29–41. Springer, Cham.

[270] Sluban, B., Smailović, J., Battiston, S., and Mozetič, I. (2015). Sentiment leaning of influential communities in social networks. *Computational Social Networks*, 2(1):1– 21.

[271] Sluban, B., Smailović, J., and Mozetič, I. (2016). Understanding financial news with multi-layer network analysis. In Battiston, S., De Pellegrini, F., Caldarelli, G., and Merelli, E. (eds), *Proceedings of ECCS 2014*, pp 193–207. Springer, Cham.

[272] Smailović, J., Grčar, M., Lavrač, N., and Žnidaršič, M. (2014). Stream-based active learning for sentiment analysis in the financial domain. *Information Sciences*, 285:181–203.

[273] Smailovič, J., Kranjc, J., Grčar, M., Žnidaršič, M., and Mozetič, I. (2015). Monitoring the Twitter sentiment during the Bulgarian elections. In Gaussier, E., Cao, L., Gallinari, P., Kwok, J., Pasi, G., and Zaiane, O. (eds), *Proceedings of the 2015 IEEE International Conference on Data Science and Advanced Analytics, DSAA 2015*, pp. 1–10. IEEE, Piscataway.

[274] Solá, L., Romance, M., Criado, R., Flores, J., del Amo, A. G., and Boccaletti, S. (2013). Eigenvector centrality of nodes in multiplex networks. *Chaos*, 23(3):33131.

[275] Solé-Ribalta, A., De Domenico, M., Gómez, S., and Arenas, A. (2014). Centrality rankings in multiplex networks. In *Proceedings of the 2014 ACM Conference on Web Science*, pp. 149–55. ACM, New York.

[276] Solé-Ribalta, A., De Domenico, M., Gómez, S., and Arenas, A. (2016). Random walk centrality in interconnected multilayer networks. *Physica D*, 323–4:73–9.

[277] Solé-Ribalta, A., De Domenico, M., Kouvaris, N. E., Díaz-Guilera, A., Gómez, S., and Arenas, A. (2013). Spectral properties of the Laplacian of multiplex networks. *Physical Review E*, 88(3):032807.

[278] Solé-Ribalta, A., Gómez, S., and Arenas, A. (2016). Congestion induced by the structure of multiplex networks. *Physical Review Letters*, 116(10):108701.

[279] Song, C., Wang, D., and Barabási, A.-L. (2013). Connections between human dynamics and network science. arXiv:1209.1411.

[280] Sowa, J. F. (1991). *Principles of Semantic Networks: Explorations in Representation of Knowledge*. Morgan Kaufmann Series in Representation and Reasoning. Morgan Kaufmann, San Mateo.

[281] Squartini, T., and Garlaschelli, D. (2011). Analytical maximum-likelihood method to detect patterns in real networks. *New Journal of Physics*, 13(8):083001.

[282] Squartini, T., Mastrandrea, R., and Garlaschelli, D. (2015). Unbiased sampling of network ensembles. *New Journal of Physics*, 17(2):23052.

[283] Starnini, M., Machens, A., Cattuto, C., Barrat, A., and Pastor-Satorras, R. (2013). Immunization strategies for epidemic processes in time-varying contact networks. *Journal of Theoretical Biology*, 337:89–100.

[284] Stehlé, J., Voirin, N., Barrat, A., Cattuto, C., Colizza, V., Isella, L., Regis, C., Pinton, J.-F., Khanafer, N., den Broeck, W., et al. (2011). Simulation of an SEIR infectious disease model on the dynamic contact network of conference attendees. *BMC Medicine*, 9(1):87.

[285] Stopczynski, A., Sekara, V., Sapiezynski, P., Cuttone, A., Madsen, M. M., Larsen, J. E., and Lehmann, S. (2014). Measuring large-scale social networks with high resolution. *PLOS ONE*, 9(4):e95978.

[286] Stumpf, M. P. H., Wiuf, C., and May, R. M. (2005). Subnets of scale-free networks are not scale-free: Sampling properties of networks. *Proceedings of the National Academy of Sciences of the United States of America*, 102(12):4221–4.

[287] Su, H.-N., and Lee, P.-C. (2010). Mapping knowledge structure by keyword co-occurrence: A first look at journal papers in Technology Foresight. *Scientometrics*, 85(1):65–79.

[288] Tacchella, A., Cristelli, M., Caldarelli, G., Gabrielli, A., and Pietronero, L. (2012). A new metrics for countries' fitness and products' complexity. *Scientific Reports*, 2:723.

[289] Takemoto, K., Oosawa, C., and Akutsu, T. (2007). Structure of $n$-clique networks embedded in a complex network. *Physica A*, 380:665–72.

[290] Tetali, P. (1991). Random walks and the effective resistance of networks. *Journal of Theoretical Probability*, 4(1):101–9.

[291] Tetlock, P. C., Saar-Tsechansky, M., and Macskassy, S. (2008). More than words: Quantifying language to measure firms' fundamentals. *Journal of Finance*, 63(3):1437–67.
[292] The Economist Intelligence Unit. (2013). *Global Liveability Ranking and Report August 2013*. Available at http://www.eiu.com/public/topical_report.aspx?campaignid=Liveability2013 (accessed May 31, 2018).
[293] Thébault, E., and Fontaine, C. (2010). Stability of ecological communities and the architecture of mutualistic and trophic networks. *Science*, 329(5993):853–6.
[294] Timár, G., Dorogovtsev, S. N., and Mendes, J. F. F. (2016). Scale-free networks with exponent one. *Physical Review E*, 94(2):22302.
[295] Török, J., Murase, Y., Jo, H.-H., Kertész, J., and Kaski, K. (2016). What does Big Data tell? Sampling the social network by communication channels. *Physical Review E*, 94:052319.
[296] Traag, V. A., Reinanda, R., and van Klinken, G. (2015). Elite co-occurrence in the media. *Asian Journal of Social Science*, 43(5):588–612.
[297] Treviño S., III, Nyberg, A., Del Genio, C. I., and Bassler, K. E. (2015). Fast and accurate determination of modularity and its effect size. *Journal of Statistical Mechanics*, 2015(2):P02003.
[298] Ugander, J., Karrer, B., Backstrom, L., and Marlow, C. (2011). The anatomy of the Facebook social graph. arXiv:1111.4503
[299] Ulam, S. M. (1960). *A Collection of Mathematical Problems*. Interscience Tracts in Pure and Applied Mathematics. Interscience Publishers, New York, London.
[300] Vanhems, P., Barrat, A., Cattuto, C., Pinton, J.-F., Khanafer, N., Régis, C., Kim, B.-A., Comte, B., Voirin, N., Pinton, J.-F., et al. (2013). Estimating potential infection transmission routes in hospital wards using wearable proximity sensors. *PLOS ONE*, 8(9), e73970.
[301] Varshney, L. R., Chen, B. L., Paniagua, E., Hall, D. H., and Chklovskii, D. B. (2011). Structural properties of the *Caenorhabditis elegans* neuronal network. *PLoS Computational Biology*, 7(2):e1001066.
[302] Vázquez, A. (2003). Growing network with local rules: Preferential attachment, clustering hierarchy, and degree correlations. *Physics Review E*, 67(5):56104.
[303] Vázquez, A., Dobrin, R., Sergi, D., Eckmann, J.-P., Oltvai, Z. N., and Barabási, A.-L. (2004). The topological relationship between the large-scale attributes and local interaction patterns of complex networks. *Proceedings of the National Academy of Sciences of the United States of America*, 101(52):17940–5.
[304] Viswanathan, G., Buldyrev, S. V., Havlin, S., Da Luz, M., Raposo, E., and Stanley, H. E. (1999). Optimizing the success of random searches. *Nature*, 401(6756):911–14.
[305] Vitali, S., Glattfelder, J. B., and Battiston, S. (2011). The network of global corporate control. *PLOS ONE*, 6(10):e25995.
[306] Wall, R. S. (2009). *NETSCAPE: Cities and Global Corporate Networks*. Erasmus University Rotterdam, Erasmus Research Institute of Management, Rotterdam.
[307] Wasserman, S., and Faust, K. (1994). *Social Network Analysis: Methods and Applications*. Structural Analysis in the Social Sciences, volume 8. Cambridge University Press, Cambridge.
[308] Watts, D. and Strogatz, S. (1998). Collective dynamics of "small-world" networks. *Nature*, 393(6684):440–2.
[309] Welch, B. L. (1947). The generalization of "Student's" problem when several different population variances are involved. *Biometrika*, 34(1–2):28–35.
[310] Wilkinson, D. M., and Huberman, B. A. (2004). A method for finding communities of related genes. *Proceedings of the National Academy of Sciences of the United States of America*, 101(Suppl. 1):5241–8.

[311] Wilson, R. J. (1972). *Introduction to Graph Theory*, volume 111. Academic Press, New York.

[312] Wolfrum, M. (2012). The Turing bifurcation in network systems: Collective patterns and single differentiated nodes. *Physica D*, 241(16):1351– 7.

[313] Yang, S.-J. (2005). Exploring complex networks by walking on them. *Physical Review E*, 71(1):16107.

[314] Yaveroğlu, Ö. N., Malod-Dognin, N., Davis, D., Levnajic, Z., Janjic, V., Karapandza, R., Stojmirovic, A., and Pržulj, N. (2014). Revealing the hidden language of complex networks. *Science Reports*, 4:4547.

[315] Zhao, K., Stehlé, J., Bianconi, G., and Barrat, A. (2011). Social network dynamics of face-to-face interactions. *Physical Review E*, 83(5):056109.

[316] Zlatic, V., Bianconi, G., Díaz-Guilera, A., Garlaschelli, D., Rao, F., and Caldarelli, G. (2009). On the rich-club effect in dense and weighted networks. *European Physical Journal B*, 67(3):271–5.

[317] Zlatić, V., Ghoshal, G., and Caldarelli, G. (2009). Hypergraph topological quantities for tagged social networks. *Physical Review E*, 80(3):36118.

[318] Zollo, F., Novak, P. K., Del Vicario, M., Bessi, A., Mozetič, I., Scala, A., Caldarelli, G., and Quattrociocchi, W. (2015). Emotional dynamics in the age of misinformation. *PLOS ONE*, 10(9):e0138740.

# Index